MICHELIN ILLUSTRATED GUIDES
TO THE BATTLEFIELDS (1914-1918)

# THE AMERICANS

## IN THE

# GREAT WAR

## VOLUME II.

### THE BATTLE OF SAINT MIHIEL

(ST MIHIEL, PONT-à-MOUSSON, METZ)

MICHELIN & Cie., CLERMONT-FERRAND

MICHELIN TYRE Co. Ltd., 81 Fulham Road, LONDON, S. W.

MICHELIN TIRE Co., MILLTOWN, N. J., U. S. A.

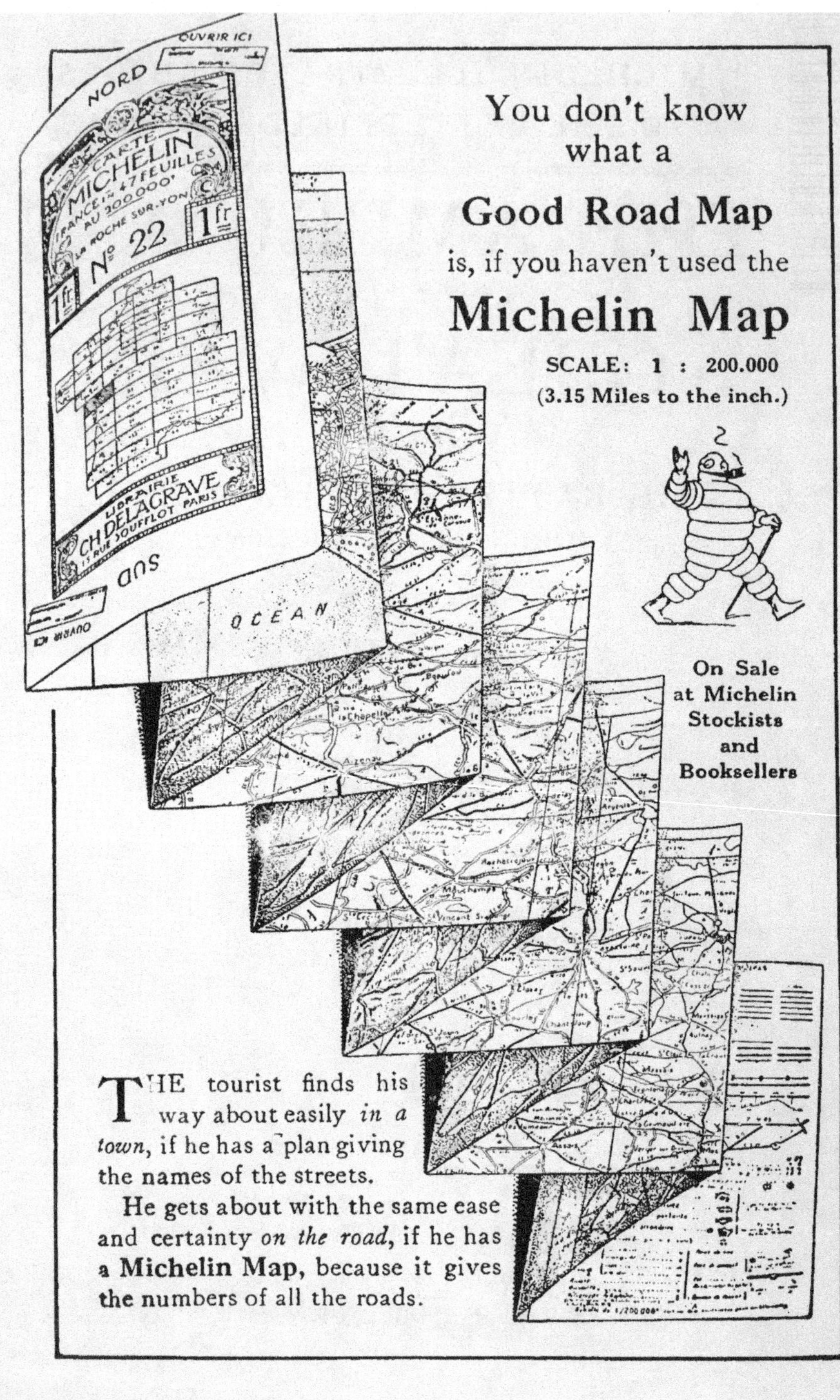


You don't know
what a

# Good Road Map

is, if you haven't used the

# Michelin Map

SCALE: 1 : 200.000
(3.15 Miles to the inch.)

On Sale
at Michelin
Stockists
and
Booksellers

THE tourist finds his way about easily *in a town*, if he has a plan giving the names of the streets.

He gets about with the same ease and certainty *on the road*, if he has **a Michelin Map**, because it gives the numbers of all the roads.

# The Michelin Wheel

BEST of all detachable wheels
because the least complicated

### *Elegant*

It embellishes even the finest coachwork.

### *Simple*

It is detachable at the hub and fixed by six bolts only.

### *Strong*

The only wheel which held out on all fronts during the war.

### *Practical*

Can be replaced in 3 minutes by *anybody* and cleaned still quicker.

It prolongs the life of tires by cooling them.

**AND THE CHEAPEST**

# THE "TOURING CLUB DE FRANCE"

## WHAT IS IT?  WHAT ARE ITS USES?

The "Touring Club de France" (founded in 1890), is at the present time the largest Touring Association in the whole world.  Its principal aim is to introduce France—admirable country and one of the loveliest on earth—to French people themselves and to foreigners.

It seeks to develop travel in all its forms—on foot, on horseback, on bicycle, in carriage, motor, yacht or railway, and soon in aeroplane.

Every member of the Association receives a badge and an identity ticket free of charge, also the "Revue Mensuelle" every month.

Members have also the benefit of special prices in a certain number of affiliated hotels; and this holds good for the purchase of guide-books and Staff (Etat-major) maps, as well as those of the "Ministère de l'Interieur," the T. C. F., etc. They may insert notices regarding the sale or purchase of traveling requisites in the "Revue" (1 fr. per line).  The "Comité des Contentieux" is ready to give them counsel with regard to travelling, and 3,000 delegates in all the principal towns are able to give them advice and information about the curiosities of art or of nature in the neighborhood, as well as concerning the roads, hotels, motor-agents, garages, etc.

Members are accorded free passage across the frontier for a bicycle or motor-bicycle.  For a motor-car the Association gives a "Triptyque" ensuring free passage through the "douane," etc.

---

**ONE TRAVELS BEST IN FRANCE WHEN A MEMBER OF THE "TOURING CLUB DE FRANCE"**

# THE AMERICANS

## IN THE

# GREAT WAR

## VOLUME II.

# THE BATTLE OF ST. MIHIEL

### (ST. MIHIEL, PONT-À-MOUSSON, METZ.)

*Published by*

Michelin & Cie, Clermont-Ferrand, (France)

Copyright, 1920, by Michelin & Cie

# FOREWORD

## THE ST. MIHIEL VICTORY. SEPTEMBER. 1918

The world already knows of the undying glory achieved in the Great War by the American Soldiers, but perhaps less is known about the historic ground over which they fought.

The purpose of the present volume is more to describe, for the benefit of the tourist, that section of France where the battle of Saint-Mihiel raged, than to dwell on the splendid achievements of the brave troops from across the seas, who took that ancient stronghold, and thus opened the way to Metz.

At the same time it is fitting to remind the reader that at Saint-Mihiel the Americans liberated over 150 square miles of French territory ; took over 15,000 German prisoners, and captured upwards of 200 guns.

President Poincaré, in a message to President Wilson, expressed in the following words, the feelings of France regarding the glorious achievements of the American troops : " I congratulate you, Mr. President, on a victory which has been completed so brilliantly. General Pershing's magnificent divisions have just liberated with admirable dash, cities and villages of Lorraine which have been groaning for years under the yoke of the enemy. I express the warmest thanks of France to the people of the United States."

Marshal Foch, also, expressed the greatest possible admiration for the way the American troops fought their way to the great victory at Saint-Mihiel. In describing the battle Marshal Foch said : " This was where the Americans for the first time showed their worth. This is where we were able to judge of these admirable soldiers, strong in body and valiant in soul. In one swoop they reduced the famous salient, which during so long we did not know how to approach."

In closing this brief introduction the publisher wishes to say that it would have been an easy matter to fill the pages following with many high-sounding phrases and verbose descriptions, but it has been thought better to adhere to the facts (they speak for themselves), and to furnish the tourist as briefly as possible with an historically correct account of the great victory of Saint-Mihiel.

# AMERICAN FORCES ENGAGED

## FIRST ARMY CORPS

### Major-General Hunter Liggett

comprising the

| | |
|---|---|
| 82nd Division | Major-General W. P. Burnham |
| 90th    „ | „    „    Henry T. Allen. |
| 5th    „ | „    „    John McMahon. |
| 2nd    „ | „    „    John A. Le Jeune. |

MAJOR-GENERAL HUNTER LIGGETT
*Commanding the 1st Army Corps.*

# FOURTH ARMY CORPS

## Major-General Joseph T. Dickman

comprising the

| | | |
|---|---|---|
| 89th Division | .. .. .. .. .. | Brig.-General Frank L. Winn. |
| 42nd ,, | .. .. . .. .. .. | Major-General C. A. Flagler. |
| 1st ,, | .. .. .. .. .. .. .. | "        "    E. F. McGlachlin. |
| 3rd ,, (Res.) | .. .. .. .. .. | "        "    B. B. Buck. |

MAJOR-GENERAL JOSEPH T. DICKMAN
*Commanding the 4th Army Corps.*

# FIFTH ARMY CORPS

## Major-General George H. Cameron

comprising the

26th Division  .. .. .. .. .. .. ..  Major-General Harry C. Hale.
4th     „      .. .. .. .. .. .. ..      „      „   Mark L. Hersey.

**MAJOR-GENERAL GEORGE H. CAMERON**
*Commanding the 5th Army Corps.*

MAJ.-GEN.
J. A. LE JEUNE,
*2nd Inf. Divn.*

MAJ.-GEN. E. F.
McGLACHLIN,
JR.
*1st Inf. Divn.*

MAJOR-GENERAL HENRY T. ALLEN
*90th Inf. Divn.*

MAJ.-GEN. C. A. F. FLAGLER
*42nd Inf. Divn.*

MAJ.-GEN. M. L. HERSEY
*4th Inf. Divn.*

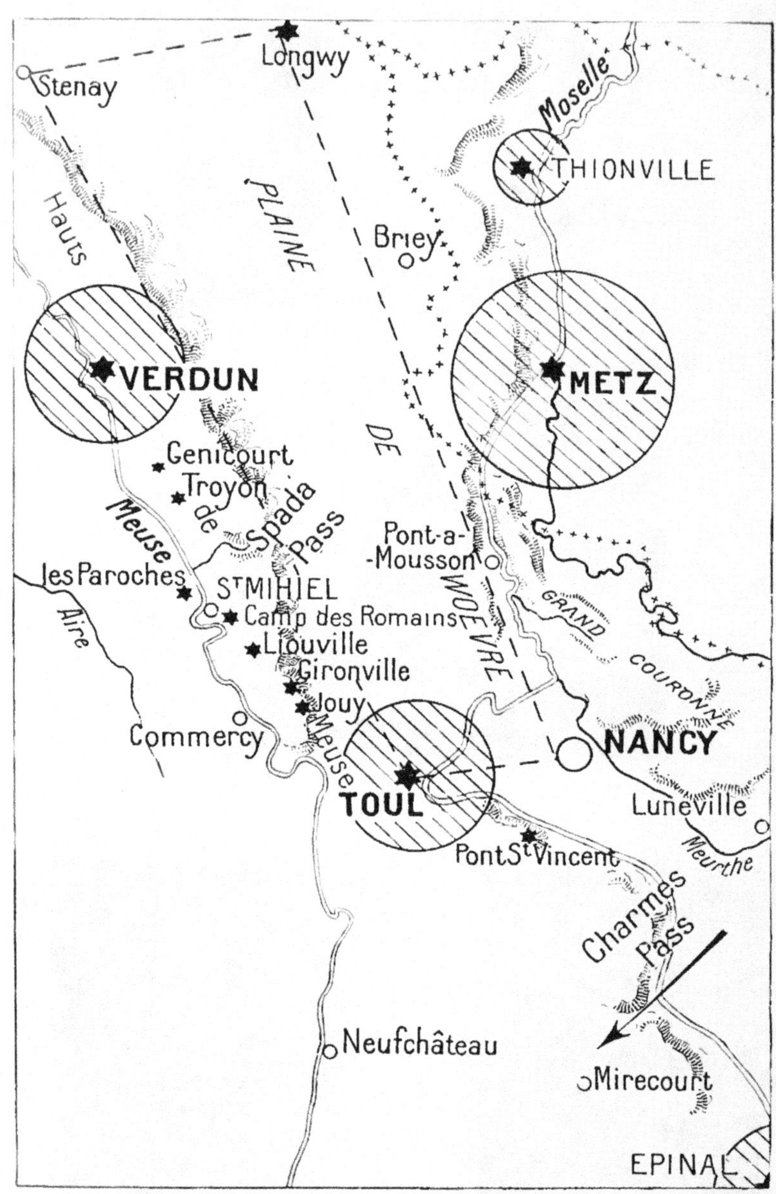


**THE FRONTIER IN 1914 AND DEFENCES OF THE MEUSE**

*In constructing these defences, Gen. Séré de Rivières' plans provided for the concentration of the French Armies to the west of the Meuse, the bridges being within range of the guns of the forts on the Meuse Heights. The quadrilateral formed by Woevre Plain was open to the enemy.*

# THE MILITARY OPERATIONS IN THE ST. MIHIEL SALIENT (1914-1918).

## The Frontier in 1914

*(See map, p. 8.)*

If we look at a map of the Franco-German frontier of 1914, between Nancy and Verdun, it will be seen that two rivers—the Meuse and the Moselle—ran parallel with the frontier, forming a double line of defences. The Moselle is protected by the hills of that name, and the Meuse by the Heights of Meuse, the eastern side of which, facing Germany, consists of a series of steep cliffs.

When, in 1875, General Séré de Rivières was instructed to fortify this frontier, his plans provided for the construction of a line of forts along the Meuse Heights, capable of holding the bridges across the Meuse under gunfire, and thus enable the French Armies to concentrate behind the river near Neufchâteau. The three northern forts, therefore, faced the Meuse ; the southern forts, viz., Gironville and Lionville, commanded both the Meuse and the Woëvre.

The drawback to this plan was that the vast Woëvre Plain lying between Stenay, Longwy, Toul and Nancy, would be sacrificed in the event of a surprise attack. The importance of this possible loss was made all the greater by the discovery of coalfields in the Briey district. It was therefore decided that a number of battalions of Chasseurs should be garrisoned in the Woëvre towns. Moreover, the passing of the Three Years' Military Service Bill made it possible to increase considerably the number of covering troops. In 1914, the Plan of Concentration provided for the grouping of the French Third Army in the Woëvre Plain. However, no permanent defences were erected. The fortress of Longwy, being isolated and of little military value, could not give effective protection.

The German Government had on several occasions given the French Government to understand that they would disapprove the erection of fortifications in Woëvre. On the other hand, the Germans unceasingly strengthened their own frontier from Metz to Thionville, increasing the perimeter of the entrenched camp of Metz from 25 to 90 kilometres and erecting ten new forts. All the attacks against the Meuse Heights started from this vast entrenched camp, which, for four years, also furnished the German lines of St. Mihiel with troops.

September 7, 1914, and the following days were particularly anxious ones for General Sarrail's army which, resting as it did on Verdun, was to form the pivot of Joffre's famous manœuvre *(see the Michelin Illustrated Guide : "The Battlefields of the Marne, 1914")*.

## HOW THE ST. MIHIEL SALIENT WAS FORMED

### First Attempt during the Battle of the Marne

*(See map, p. 10.)*

A furious frontal attack was made on this army by the ex-Crown Prince of Germany, while at the same time it was taken in the rear, on the Meuse Heights, by the Bavarian Crown Prince. Had the latter succeeded in crossing the Meuse, Verdun would have become untenable; General Sarrail would have been forced to retreat southwards, and, as in this gigantic battle of the Marne all the armies were interdependent, such a withdrawal would have been felt all along the line, and Joffre's plans for a strategical recovery would have failed.

On September 8, the Germans bombarded the Fort of Troyon. The Governor of Verdun telegraphed to the officer in command of the fort that victory depended upon his resistance, and requested him to hold out " indefinitely." As a precautionary measure, General Sarrail ordered several of the bridges across the Meuse to be destroyed.

On September 9, the fort's guns were put out of action, but the defenders repulsed several assaults. Génicourt Fort was next bombarded.

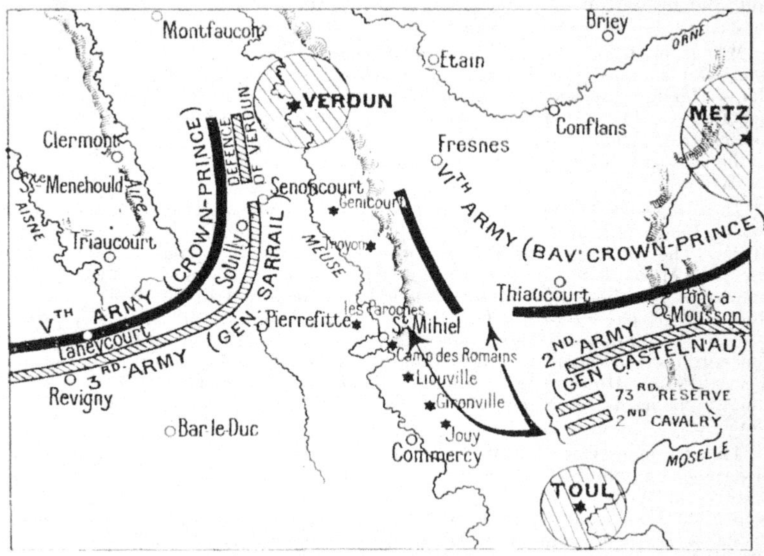


THE FIRST GERMAN ATTACKS AGAINST THE MEUSE HEIGHTS

*While the Battle of the Marne was raging, the Germans attempted in vain to capture the Meuse Heights, in order to take Gen. Sarrail's Army—the pivot of Joffre's manoeuvre—in the rear.*

On the 10th, the forts were still holding out, although deluged with shells. Meanwhile the German infantry advanced towards St. Mihiel.

However, the Battle of the Marne had now been won on the left wing, and the German retreat, which was to extend as far as the Verdun—St. Mihiel district, had begun.

General de Castelnau despatched the 73rd Reserve and 2nd Cavalry Divisions of his army to Troyon, and the fort was relieved on September 13. The mobile defence forces of Verdun pursued the retreating Germans across the Meuse and established themselves to the east of the town, while General Sarrail's army advanced towards the north and west.

The German plan had completely failed.

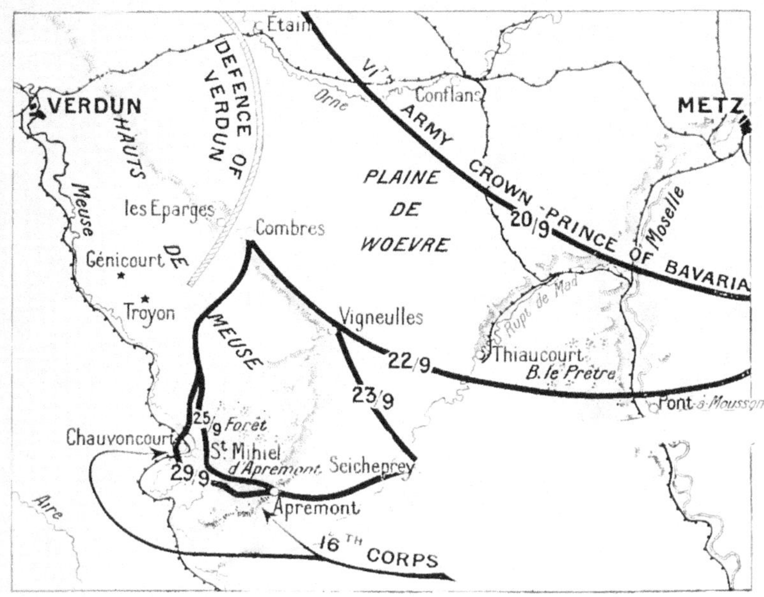



THE FORMATION OF THE ST. MIHIEL SALIENT (*Sept.* 20-29, 1914)

Their set-back at Troyon did not prevent the Germans reforming, and they attacked the Meuse Heights again on September 20, in an endeavour to outflank Verdun from the south. Four army corps under General von Strantz, starting from Metz, advanced rapidly on the 22nd to the Combres-Vigneulles-Thiaumont line, and began a methodical bombardment of the forts on the Meuse Heights. These were soon pounded into shapeless heaps of *débris*, but the gallant defenders still held on and repulsed all assaults.

On the 23rd, the enemy advanced to Seicheprey. The mobile defence forces of the region, greatly outnumbered (two or three to one), formed only a very thin line, the depth of which steadily decreased as it extended beyond Verdun to the south.

On September 24, the German attacks were renewed with increased fury. On the 25th, they succeeded in gaining a footing on the Meuse Heights near Vigneulles, whence they advanced to St. Mihiel, which they entered without, however, crossing the Meuse. At this point the river was only defended by one battalion of Territorials, and the Germans were able to cross on the 26th, after which they began to advance towards the valley of the Aire, in the direction of Verdun. The situation was critical. The 16th Corps from Nancy met and defeated the German forces, and obliged them to fall back in disorder on the suburbs of St. Mihiel, but were unable to force them back across the river. On September 29, the front line ran through Combres, Chauvoncourt, Apremont and Seicheprey.

The salient had been made.

THE DESTROYED BRIDGE AT ST. MIHIEL
*In the background:* Temporary foot-bridge built by the Allies in Sept., 1918.

## THE ST. MIHIEL SALIENT—Oct., 1914, to Sept., 1918

*(See map. p. 13.)*

From November 17-20 the French endeavoured to drive the Germans from the bridgehead which the enemy held at Chauvoncourt, opposite St. Mihiel. In a spirited attack they drove the Germans from the suburb and barracks of Chauvoncourt. However, the latter had been mined, the Germans, taking

FORTIFIED STREET IN FEY-EN-HAYE (1915)

advantage of the confusion caused by the explosion, counter-attacked and reoccupied Chauvoncourt.

This was the last attack made at the point of the salient. Only local fighting of extreme violence now took place in Apremont Forest, the result of which was, the French prevented the Germans extending the salient.

French offensives were launched against the northern and southern sides of the salient, at Eparges and Prêtre Wood, in the hope of narrowing the salient and forcing the Germans to evacuate it. Eparges Crest was conquered after more than two months of the fiercest possible fighting, ending on April 9, 1915.

However, these local actions were insufficient, and little by little the line became fixed. Both sides entrenched themselves and bombarded each other unceasingly, while the sappers carried out long and strenuous mining operations. Attacks were henceforth confined to small local objectives: a wood,

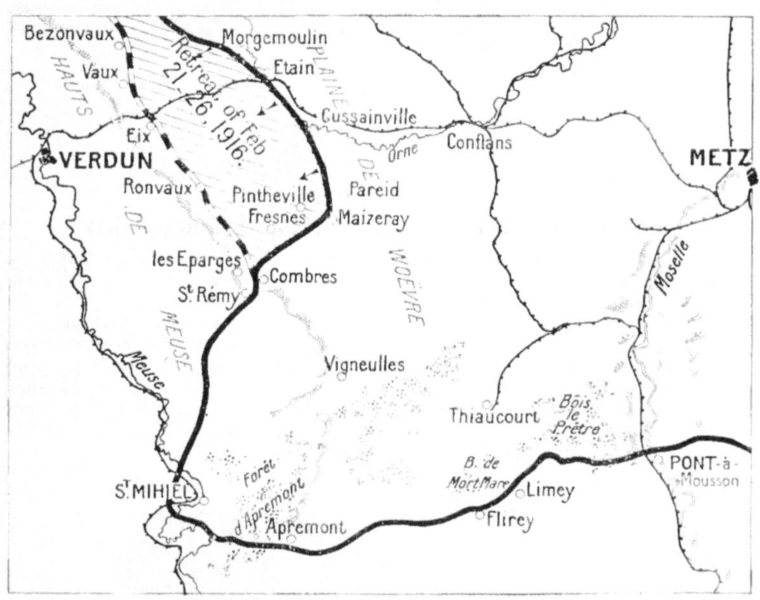

ST. MIHIEL SALIENT, FROM OCT., 1914, TO SEPT., 1918

house, bridge or crater, and it required the great American offensive of September, 1918, to flatten out this salient which, for four years, had formed a huge "pocket" inside the French lines.

### The Salient during the Battle of Verdun

The German offensive, which began on February 21, 1916, caused a slight withdrawal along the whole of the French Verdun-Nancy line (*see the Michelin Illustrated Guide: "Verdun"*). The French line was withdrawn behind Fresnes, passing thence round Eparges Crest, which formed a hinge.

After the French counter-offensive of July-September, 1917, which disengaged Verdun and the immediate vicinity, their positions were further improved by a series of local operations at Eparges and around Pont-à-Mousson.

ON BEAUMONT HEIGHTS

*Gen. F. E. Bamford, commanding the American 2nd Brigade,
watching the advance of his troops before Beaumont, Sept. 12, 1918.*

## THE AMERICAN OFFENSIVE OF SEPTEMBER, 1918

It has been seen in the *Michelin Illustrated Guide :* "The Americans
in the Great War," Vol. I., that the 1st and 3rd American Corps, under the
respective commands of Major-Generals Liggett and Bullard, reached the
Vesle at the beginning of August, 1918. General Pershing's intention at
that time was to use these two army corps to form the American First Army
which, under his personal command, was to relieve the French 6th Army
(General Degoutte). However, the Germans having given proof of their
intention to defend the Vesle line at all cost, Marshal Foch decided to attack
at another point of the front, and entrusted the task of flattening out the
salient to the American Army.

FLIREY VILLAGE (*Sept.* 14, 1918)

*American Sappers pulling down the walls of the ruined houses
to fill in the German trenches across the roads in the salient.*

This operation had already been carefully studied by the American Staff, for it was in this region that the first American divisions were trained in active warfare.

The 1st Division was holding the sector extending from Ailly Wood to Mortmare Wood, when it was relieved by the 26th Division on April 2, 1918, and despatched to the Somme, where it covered itself with glory by the capture of Cantigny. On April 20, the 26th Division withstood a powerful surprise attack at Seicheprey, where, after losing part of the village, it succeeded in fully re-establishing its front. On July 10, it was sent from the Woëvre district to take part in the Battle of the Ourcq.

From January, 1918, the 2nd Division held that part of the front lying between Eparges and Spada Pass, where it received a thorough training, the effects of which the Germans were destined to feel around Château-Thierry in June, 1918.

On August 30, General Pershing took over the command of the First Army, with Headquarters at Ligny-en-Barrois. At that time, the front line of the salient ran as follows: from Eparges Crest it descended in an almost straight line to St. Mihiel, along the Meuse Heights ; passing thence round St. Mihiel, the great bend in the Meuse and the Camp des Romains, it described a vast semicircle ; then turning sharply eastwards, it proceeded towards Pont-à-Mousson, passing through the woods of Apremont, Ailly, Mortmare and Le Prêtre.

The total length of the salient front was about 65 km., and its width along the German lines between Eparges and Regniéville (near Prêtre Wood) about 39 km. It penetrated the French lines to a maximum depth of 22 km.

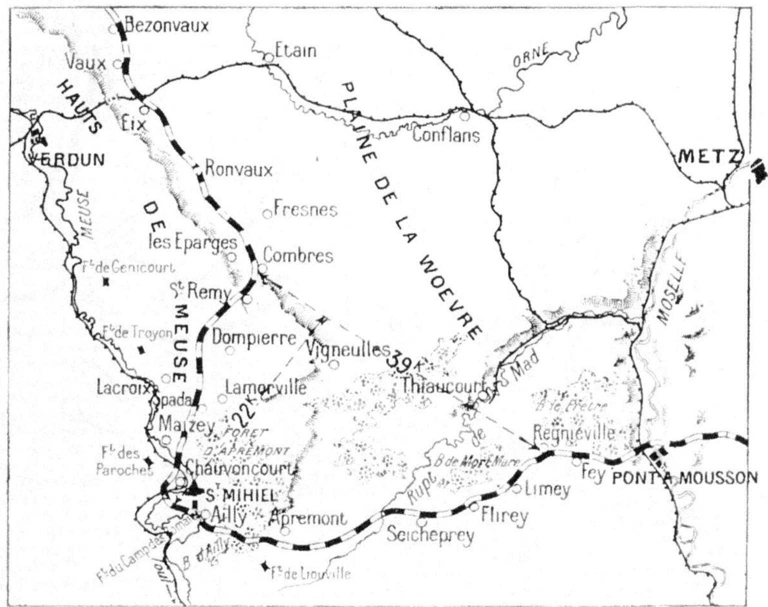

ST. MIHIEL SALIENT PRIOR TO THE OFFENSIVE OF SEPT., 1918

*It measured 39 km. across its greatest width, 22 km. in depth, and about 65 km. along its front.*

Since 1916 this important salient had been fairly quiet, and beyond intermittent bombardments—which showed that the lines on both sides were defended and that the artillery was on the alert—and a few local attacks, the communiqués had nothing to report. This salient, however, greatly hampered the French lines of communication, cutting as it did the railway between Verdun and Toul. This line, which runs as far as Epinal and Belfort, linked up these four great eastern fortresses before the war.

## The Defences of the Salient

### *(See map below.)*

Through aerial observations and prisoners taken during raids, the American High Command knew that the enemy possessed several lines of defences, one behind the other, in the salient, and that beyond the first line of trenches facing the front was a second line known as the Schroeter Zone, which formed a second salient about 5 km. within the first. This line began northeast of Eparges, and went southwards across the Meuse Heights, then descending eastwards near Varvinay as far as Buxières, afterwards passing behind the deep valley of the Rupt-de-Mad, and lastly going in a north-easterly direction through Nonsard, Lamarche, Beney and Xammes, where it joined up with the Michel line. The latter formed part of the system of defences known as the *Hindenburg Line or Kriemhilde Position*—considered impregnable by the Germans, and of which they said to the Allies: "Thus far, and no further"—and it was there that the final enemy stand in the salient was to be made.

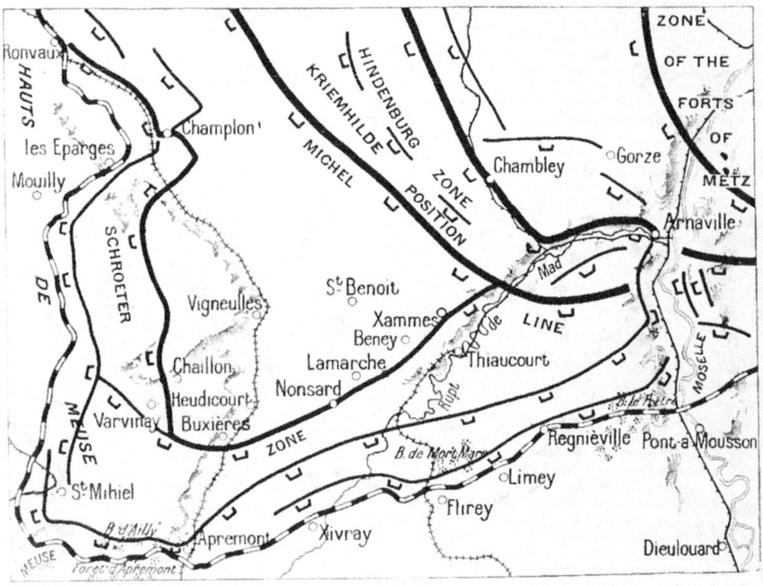


THE GERMAN DEFENCE WORKS IN THE SALIENT
*The German lines of defence extended in échelons over the whole depth of the salient, and rested on the zone of the advanced forts of Metz.*

## The Opposing Forces

*(See map below.)*

Lieutenant-General Fuchs, Commander of the German forces in the salient, had eight divisions in the line and five divisions of reserves.

These divisions formed part of the forces of General von Gallwitz, commanding the army group, and it was he who really opposed the Americans.

On this front General Pershing had four army corps disposed as follows:—

The 1st Corps, comprising the 82nd, 90th, 5th and 2nd Divisions, commanded by Major-General Hunter Liggett, operated from Clémery, east of the Moselle, to Limey.

The 4th Corps, consisting of the 89th, 42nd and 1st Divisions, commanded by Major-General Joseph T. Dickman, operated from Limey to Xivray.

To these two Corps was assigned the task of carrying out the main attack,

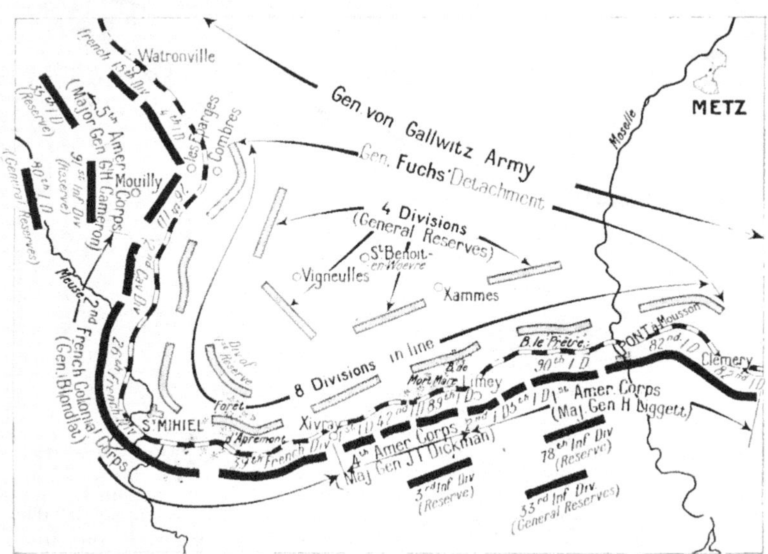


THE OPPOSING FORCES AT THE BEGINNING OF THE 1918 OFFENSIVE

their objective being the Vigneulles-St. Benoit-Xammes line, which was to be reached in three successive rushes.

The 5th American Corps, composed of the 26th and 4th Divisions under Major-General George H. Cameron, and supported by the French 15th Division, carried out a secondary attack from Mouilly to Watronville, the objectives being, first the capture of the crests of Eparges and Combres, then the Combres-Vigneulles line. The Corps was to join hands in the latter village with the troops engaged in the main attack.

The French 2nd Colonial Corps, first under General Blondlat and afterwards General Claudel, operated in the centre of the salient, from Xivray to Mouilly, with orders to protect the flanks of the two American attacks.

The attacking forces consisted of some 216,000 Americans and 48,000 French, in addition to the American Reserves (190,000 men), who were ready at a moment's notice to take part in the battle.

In his official report General Pershing stated that he had mustered a body of troops three times as large as General Grant's Army of the Potomac in 1864-1865.

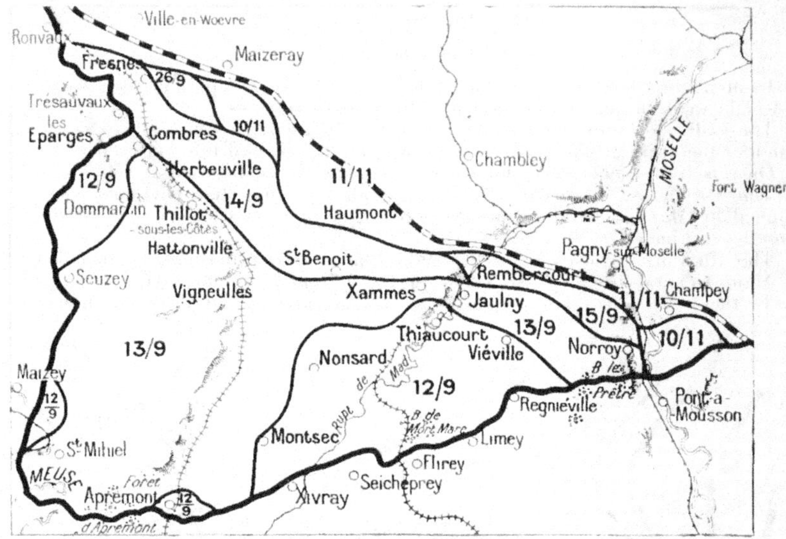


SHOWING THE AMERICAN-FRENCH ADVANCE FROM SEPT. 12 (12/9) TO<br>NOV. 11 (11/11), 1918

*Two secondary attacks on Sept. 12 held the enemy at the bottom of the salient, while the main attacks on the flanks crushed in the latter, as in the jaws of a vise. On Sept. 13, the Germans, in danger of being cut off, were forced to evacuate the salient.*

## Flattening out the Salient, Sept. 12, 1918

### (See map above.)

Despite all the precautions taken by General Pershing to ensure the secrecy of his troops' movements in the St. Mihiel sector, the Germans expected the attack, and as early as the beginning of September began to withdraw their heavy guns, and to make active preparations for the total evacuation of the salient. However, General Pershing did not give them time to do this, and ordered the attack to be made on September 12, at 5 a. m. for the 1st and 4th Corps, and at 8 a.m. for the 5th Corps.

The attack had been worked out in minute detail, and the time-table of the advance exactly laid down. Everything took place as arranged. After an artillery preparation lasting four hours, the American divisions advanced, supported by a certain number of tanks, half of them driven by Americans and the other half by Frenchmen. Accompanied by soldiers whose duty it was to cut the barbed wire, and by men armed with "bangalore torpedoes," the Americans advanced in successive waves. They soon reached the enemy trenches and fell unexpectedly on the demoralized foe in the middle of a fog.

On September 12 (12/9) the 1st Corps quickly took Thiaucourt, whilst the 4th Corps, operating on the left, advanced beyond Montsec and reached Nonsard, further north. At the point of the salient, the 2nd French Colonial Corps gradually attained the objectives assigned to it. The 2nd Cavalry Division captured more than 2,500 prisoners with a loss of only fourteen men killed and 116 wounded. At the other end, the 5th American Corps carried the crests of Eparges and Combres, repulsed a counter-attack, and quickly joined hands with the patrols of the 4th Corps at Vigneulles.

On the morning of September 13 (13/9), Generals Pershing and Pétain entered St. Mihiel. In the evening the new front line ran as follows: Herbeuville, Thillot-sous-les-Côtes, Hattonville, St. Benoit, Xammes, Jaulny, Norroy.

It was a fine victory: 16,000 prisoners, 443 guns of all calibres, and huge quantities of stores and munitions were captured, with a loss of only 7,000 killed and wounded.

The German retreat continued on September 14 and 15 (14/9 and 15/9) to the line Fresnes—Hautmont—Rembercourt.

The offensive was finished ; the jaws of the vise had closed on the salient, and the latter had disappeared. From the American advance-posts the outworks of Metz were now plainly visible, and Wagner Fort, situated in front of the town, was already under the fire of the American guns.

## German Comments on the Attack

A German report on the American attack of the St. Mihiel salient contains the following:—"*The Americans made a clever use of their machine-guns. They are stubborn in defence, and rely greatly upon this weapon, of which they have large numbers.*

"*The artillery preparation, which preceded the attack, was well carried out. The objectives were efficiently bombarded. The American gunners were able to change their targets in the minimum of time, and with great accuracy. The liaison between the infantry and the artillery was faultless. Whenever the infantry were stopped by a nest of machine-guns, they immediately fell back, and their artillery promptly shelled the nest of machine-guns.*

"*Numerous tanks were ready, but only a few actually used ; the masses of infantry alone ensured the victory.*"

COMRADES IN ARMS

MARSHAL FOCH　　　　GENERAL PERSHING

### France's Congratulations

Immediately after the first American successes in the salient, the President of the French Republic cabled his warmest congratulations to President Wilson for the victory of "*General Pershing's magnificent divisions, fraternally seconded by French troops.*"

The praise was well deserved, as in two days the Americans had liberated 150 square miles of French territory which had been occupied four years by the enemy.

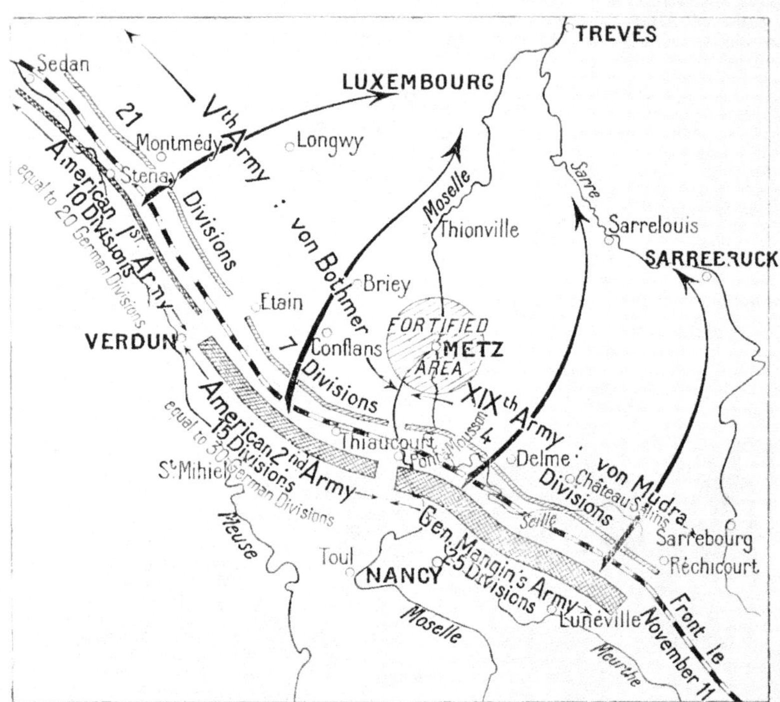


THE OPPOSING FORCES ON ARMISTICE DAY

*A crushing offensive was on the eve of being launched. The enemy, incapable of effectual resistance, hauled down their flag and capitulated.*

## St. Mihiel Front from Sept. 15 to Armistice Day

During the great Meuse-Argonne Battle, fought by General Pershing's troops after September 26, the operations on the St. Mihiel front were limited to intermittent bombardments and local attacks.

When the Armistice was signed on November 11, General Pershing was making dispositions to invest Metz by an offensive towards Longwy with the 1st Army, and towards Briey with the 2nd Army, while a detachment of six American divisions was to co-operate on the right bank of the Moselle with General Mangin's Army, in an attack on Château-Salins. Meanwhile, the Germans had already begun to evacuate Metz. The Allies' advance began on November 10 and 11, but the general capitulation of the Germans, on terms dictated by the Allies, robbed the Americans of a new and crushing victory, which would have fittingly crowned their fine success at St. Mihiel.

# A VISIT TO THE BATTLEFIELDS
## IN THREE ITINERARIES

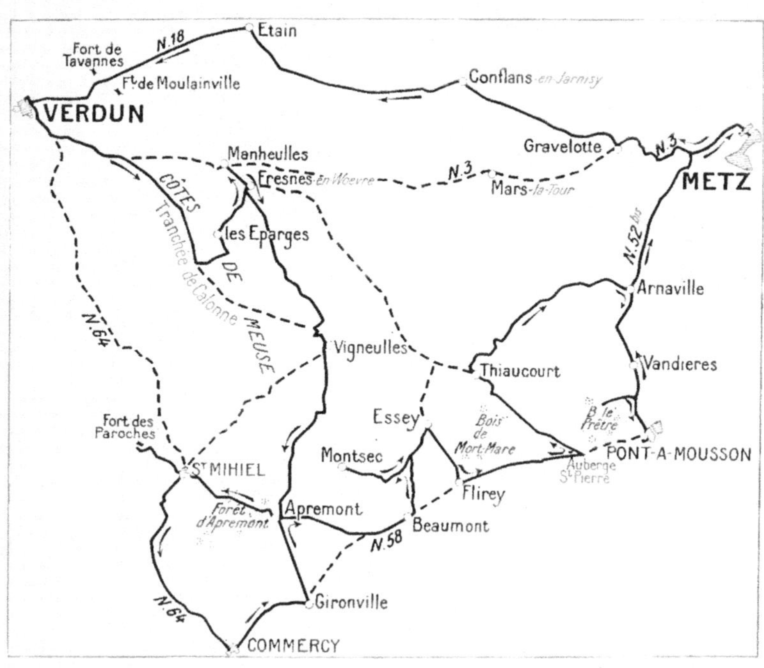


FIRST ITINERARY (p. 22)
Distance: 80 km. (See pp. 23-71)

**VERDUN to COMMERCY, via Calonne Trench, Eparges, Apremont
Forest, Ailly Wood and St. Mihiel, including
A VISIT TO ST. MIHIEL** (pp. 55-69)

SECOND ITINERARY (pp. 72-137)
Distance: 112 km. (See pp. 72-137)

**COMMERCY to METZ, via Pont-à-Mousson, including
A VISIT TO PRÊTRE WOOD** (pp. 102-119)

**A VISIT TO METZ** (120-137)

THIRD ITINERARY (pp. 138-145)
**METZ to VERDUN, via Etain** (pp. 138-145)

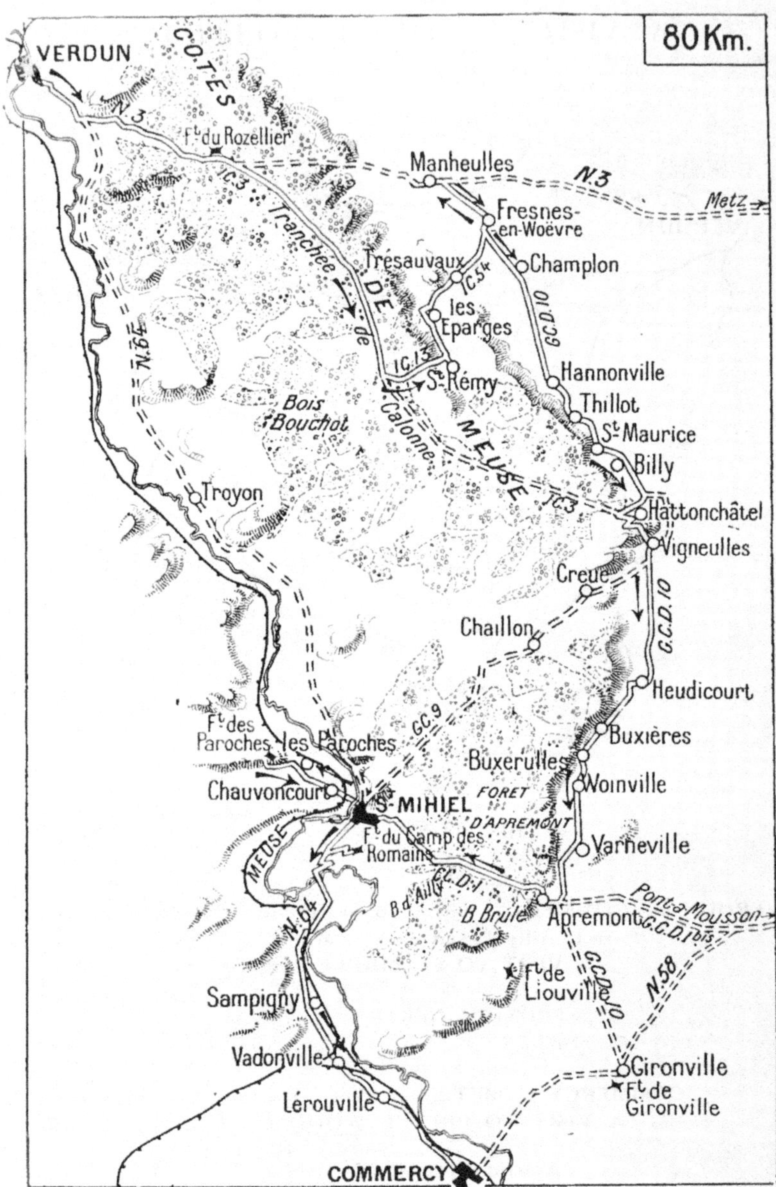


**FIRST DAY—VERDUN TO COMMERCY**

*Follow the roads indicated by the continuous black lines, in the direction of the arrows. See sheets 7 and 12 of the Michelin Touring Map.*

FAÇADE OF THE HÔTEL-DE-VILLE, VERDUN, OVERLOOKING
THE PUBLIC GARDENS

*(From the Michelin Guide: The Battle of Verdun.)*

# FIRST DAY

## FROM VERDUN TO COMMERCY

*Leave Verdun by the Rue de l'Hôtel-de-Ville, Rue St. Sauveur, Rue and Gate of St. Victor (photo below) and N. 3.*

*Ten kilometres down this road, Rozellier Fort will be seen on the left. One kilometre further on, take the strategic I.C. 3, also known as Calonne Trench, on the right.*

VERDUN—ST. VICTOR'S GATE
*(From the Michelin Guide: The Battle of Verdun.)*

**CALONNE TRENCH**

*French Post of Commandment on the left, about 200 yds. before the fork in the road to Haudiomont (see sketch map, p. 25).*

## Calonne Trench

This picturesque road enables the tourist to follow the phases of the struggle which took place in the district of Les Éparges. The road crosses in an almost straight line the whole forest of Amblonville, Bouchot Wood, and La Montagne Forest, and comes out about twelve miles further on at the Hattanchâtel cross-roads. Formerly this road was used only by poachers, gamekeepers, and shooting-parties, being a well-known haunt of game.

Calonne Trench will, in future, evoke more tragic memories. The name "Trench" might lead one to suppose that it dates from the Great War, but this is not the case. For more than a century the road, cut out of the crest of the hills, has borne this name. It was made by order of M. de Calonne, Minister of Finance under Louis XVI., to give access to his château at the foot of the Meuse hills. This château was destroyed during the Revolution.

**CALONNE TRENCH**

*The "Bouée" Post of Commandment, 1 km. after the fork, and 100 yds. in the wood on the left of the road (see sketch map, p. 25).*

CALONNE TRENCH

*French Trenches and Observation-Post on the right, before reaching the road
to Éparges (see sketch map, p. 26).*

It is said that M. de Calonne, hoping some day to entertain the King at
his château, had rose-trees planted the whole length of this road. However
that may be, it is a fact that during the War wild roses were seen in bloom
all along this forest road, at that time really a " trench " in the military sense
of the word.

The battle-front crossed Calonne Trench a little to the south-west of St.
Remy, in Bouchot Wood. Both adversaries bombarded each other and were
kept constantly on the alert by attacks and counter-attacks. In March, 1915,
5.5-in. naval guns with a range of 13,000 yards, were placed in position, to
fire over Les Éparges, behind the enemy lines.

The marines had great difficulty in bringing these heavy guns into action,
owing to the slippery, clayey soil.

Their effective bombardment ir-
ritated the Germans so much
that on April 20 they bombarded
the French lines and, four days
later, launched a massed attack
which reached the third line of
support.

The marine officers, cut off from
their base and unable to communicate with the infantry—the
telephone wires being cut—hastily
organised defences. They swept
the ground with the fire of their
heavies and some 75's brought up
by hand, which opened at fuse 0.

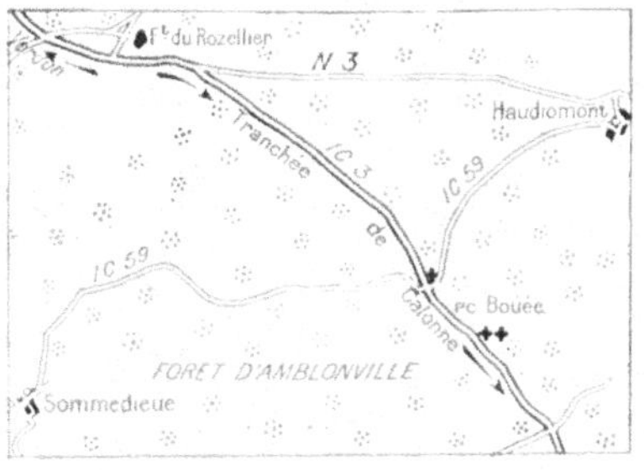

**CALONNE TRENCH**
*On the left: Road to Éparges (impracticable).*

Meanwhile the Germans continued to advance. On the 25th they were within a thousand yards of the guns, and only vestiges of the trenches and of the original barbed wire entanglements lay between them and the guns. On the 26th, while the marines were preparing a vigorous resistance, two battalions of French Chasseurs, summoned to reinforce them, crept through the brushwood and began a counter-attack. On the 27th, the firing became more distant, but the Germans re-formed and renewed the attack on May 5. At first they met with some success, but this was quickly changed by the intervention of the Moroccan Brigade and six battalions of Chasseurs, who retook in a few hours all the ground lost on April 24.

Calonne Trench enters the forest almost immediately. On both sides of the road are numerous engineer and artillery parks, ambulance stations, shelters, rail-tracks and gun-pits.

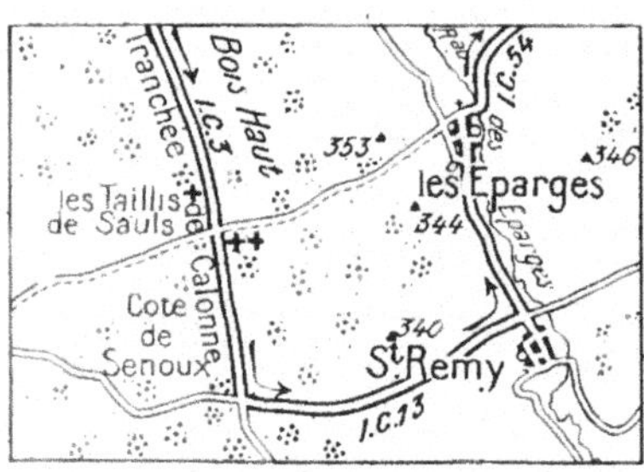


+ *Photo, p. 25.*
++ *Photo, p. 26.*

*Three kilometres from N. 3 and on the left, 200 yards before reaching the fork in I.C. 59, which leads to Haudiomont, there is a French Post of Commandment (photo, p. 24); fifty yards to the right, beyond the fork, a military cemetery; 1 km. beyond the fork, on the left of the road, a hundred yards in the wood, the "Bouée" Post of Commandment (photo, p. 24); 2 km. further on, to the left, a French military cemetery. Leave the fork of Mont-sous-les-Côtes on the left and follow the road. In the "Taillis de Sauls" the*

ROAD FROM CALONNE TRENCH TO ST. REMY
*German shelters and dug-outs.*

French first lines (trenches, shelters, dug-outs, barbed wire entanglements and observation-posts) begin ; *on the left*, a military cemetery ; *on the right*, a concrete shelter.

From this point to where the destroyed road to Eparges begins (*photo, p. 26*) the forest consists of little more than blackened shell-torn tree-stumps.

*Continue along Calonne Trench which, for 1,500 yards, crosses Senoux Hill.* Here the spectacle is appalling, especially on the site of the German trenches. *Bear to the left and take I.C. 13 towards St. Remy. It is a bad road, but with care passable. For 2½ km. it descends to the Eparges stream.*

All along this road, cut out of the left side of the hill, are concrete shelters.

RUINED CHURCH AND VILLAGE OF ST. REMY
*In the background:* Combres Crest (*right*), Eparges Crest (*left*).

THE CHOIR OF ST. REMY CHURCH

*Note the German stone and concrete Gun Shelter. The Germans bombarded Eparges from here.*

dug-outs, underground passages, German posts of commandment, and a few German graves.

*In the valley, a cross-road is reached, close to the stream. Take the road to the right to St. Remy, the ruins of which are seen in the distance.*

*Climb up to the church (German graves in the cemetery). Fine extensive view across the valley towards Combres and Eparges.*

*In the church, where the altar formerly stood, is a German shelter of stone and concrete, which concealed a big gun firing on Eparges (photo above).*

*Return to the cross-road and continue along the bottom of the valley (I.C. 54)*

EPARGES VILLAGE (*coming from St. Remy*)

*On the right:* Montgirmont Hill; *on the left:* Hures Hill.

EPARGES VILLAGE

*The Cemetery is in front of the last house on the right of the road to Tresauvaux.*

*to the village of* **Eparges.** The road crosses the original German and French front lines.

GENERAL VIEW OF EPARGES HEIGHTS, SEEN FROM MONTGIRMONT CREST

*A, The Woëvre ; B, Trench along Montgirmont Crest (the photo was taken from here) ; C, Eparges Crest ; D, Death Ravine ; E, Shelters in the sides of Eparges Crest ; F, Trench.*

*Go through the village,* of which only a few walls remain standing.  Numer-ous French defence-works, including some of concrete.

*At the last house the road turns to the right in front of* a French cemetery, *and goes towards* **Trésauvaux,** *passing between* Montgirmont Crest *on the right*

DEATH RAVINE (1915)

*and* Hures Hill *on the left (photo, p.* 28). All along the trenches, shelters and numerous graves.

*At the top of the hill* the houses of Trésauvaux *come into view. Here leave the car and climb the slopes of Montgirmont* (trenches, boyaux, etc.), *from the top of which there is a fine* panorama of Eparges on the French side (*photo, p.* 29).

It is a desolate scene. The side of the hill is full of craters and shell-holes, forming so many grey patches on the reddish earth on which no vegetation survives. The glorious crest, entirely bare, stands out against the sky. Death Ravine, where so many brave men fell in the first assault on Eparges, lies between Montgirmont (*where the tourist stands*) and Eparges.

## Eparges Spur

Eparges Spur, 1,500 yards in length and over a thousand feet high, forms the end of Woëvre Plain. Its sides are steep and slippery, while numerous springs and rivulets run down its slopes. It has been rightly called " a mountain of mud." Eparges Heights form part of a series of hills, among which are Hures, Montgirmont, Combres and St. Remy. Of these, Eparges Crest is the most important. By nature an observation-post, its possession enabled those who held it to keep all the surrounding roads under gunfire.

The Germans captured it on September 21, 1914, and immediately made several lines of trenches between the summit and the valleys. At some points, five rows of batteries, one above another, were placed, and nowhere were there less than two.

Facing Eparges Crest, the French held the brow of Montgirmont to the north, and below, the village of Eparges, only 600 yards from the German trenches. Between Montgirmont and the northern slopes of Eparges Heights, an earth track crosses the pass between the two hills. It was on the western side that, at the end of October, the French began the attack, sapping step by step, while at the same time they slipped into the woods on the north-east, which cover the side of the ravine.

From February onward, attacks and counter-attacks took place almost daily and only came to an end early in April, after the French had captured the crest. On February 17, the explosion of a mine enabled the French to enter the west sector of the enemy's first line. Attacks and counter-attacks continued for five days, during which Colonel Bacquet was mortally wounded while leading his troops. The French held the whole of the western bastion, and began to make progress towards the eastern bastion. From March 13 to 21 they renewed their attacks and captured the enemy's first line.

On March 27, a battalion of Chasseurs made a fresh advance, and on April 5 began the last great attack which was to continue day and night until the 9th.

Two regiments attacked in the rain, but the muddy ground greatly impeded their movements, and it seemed at times as if the attack would fail.

In the evening the French occupied some of the trenches, but the use of aerial torpedoes, which pulverized whole rows of men, and a massed counter-attack launched at 4.30 on the morning of the 6th, forced them to give up part of the ground gained in the first advance. On the evening of the 6th, and throughout that night, in spite of the incessant rain, the trenches were retaken and the enemy driven back foot by foot, with a loss of 100 prisoners, including several officers. The French replied to the German counter-attacks with bayonet charges or barrage fire. The communicating trenches were bombarded, levelled, or blocked up. On the 8th, two regiments of infantry and a battalion of Chasseurs made a fresh bayonet charge. At 10 o'clock the summit and the western crest were strongly held, and by midnight, after fifteen hours of strenuous, uninterrupted fighting, almost the whole of the crest was in the hands of the French.

During the night of the 8th, the relief of the troops was carried out, but the ground was so muddy that men sank into it, stumbling and slipping at every step. Fourteen hours passed in blinding rainstorms before the fresh troops were established in position. At 3 o'clock in the afternoon of the 9th the attack was resumed. The ground was full of deep holes in which men sometimes disappeared. At the moment when the eastern edge of the plateau was reached a cloud of fog descended over the crest. Firing was out of the question. The Germans counter-attacked and forced the French to retreat momentarily, but half an hour later the French retook the lost ground in a furious charge, and by 10 o'clock at night held the whole of the Eparges Heights. Only Combres Hill, threatened by the machine guns of Eparges and St. Remy, remained in the hands of the Germans.

The enemy had left nothing undone to put the position in a state of defence. Their cave-shelters contained a narrow-gauge railway, sleeping quarters, and

EPARGES IN 1915.
POST OF COMMANDMENT IN THE SIDE OF THE CREST

even an officers' club. Their relief reinforcements were concealed from the French, while their cannon and machine-guns were unceasingly turned on the muddy slopes up which the French laboriously climbed. Unwounded men were drowned in the mud, while many of the wounded could not be rescued in time from the quagmires into which they had fallen.

The victory of Eparges has been described as "a work of giants." But it was a costly victory. Most of the officers and thousands of men fell. The German losses were at least as heavy as those of the French.

*Return to the Trésauvaux road.* The village of **Trésauvaux,** the ruins of which were organised militarily by the French (*photo. p. 34*), *is reached shortly afterwards.*

EPARGES IN 1915
*Making rings at the entrance to a dug-out during a lull.*

TRÉSAUVAUX VILLAGE

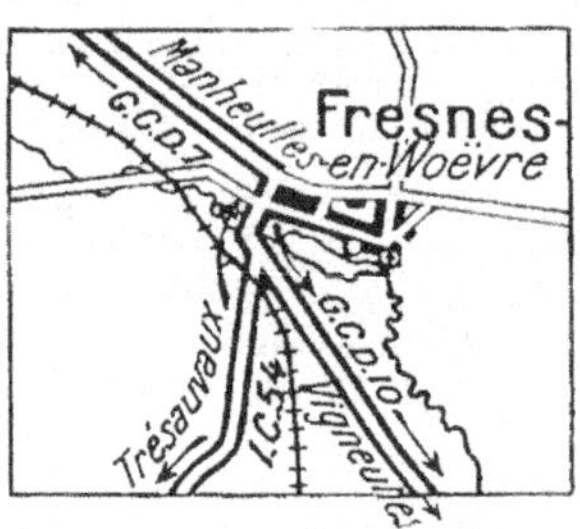



*Follow I. C. 54 to* **Fresnes-en-Woëvre,** *the* houses of which are seen in the distance. *Go through the ruined village (photos, p. 35).* The statue of General Margueritte has been severely damaged, while the church is entirely in ruins.

*Leaving Fresnes, take the Manheulles road (G.C.D. 7) on the left, which joins N. 3, 2½ kms. further on.*

*Continue as far as* **Manheulles,** *where there are numerous military works,* includ-

THE SQUARE, TRÉSAUVAUX, IN 1915

FRESNES-EN-WOËVRE CHURCH, AUG. 11, 1915

FRESNES-EN-WOËVRE. MUTILATED STATUE OF GEN. MARGUERITTE

GERMAN POST OF COMMANDMENT AT ENTRANCE TO
MANHEULLES VILLAGE (*see p. 36*)

MANHEULLES.  THE MAIN STREET

ing concrete blockhouses and a German post of commandment *established in
the first house on the right at the entrance to the village.* The ground-floor
of this house, which appears to be an absolute ruin, is lined throughout with
concrete (*photo p. 35*). Several concrete shelters have been added outside,
on the front facing Woëvre Plain.

*At the cross-roads, in the middle of the village, stands* a machine-gun blockhouse, built of concrete (*photo below*).

*From Manheulles return to Fresnes-en-Woëvre.*

*Take G.C.D. 10, which passes through the villages of* Champlon and Hannon-
ville (severely damaged), Thillot, St. Maurice *and* Billy. *Leave the village of
Viéville on the right and, 500 yards further on, take the road which leads up a
steep slope to* **Hattonchâtel.**

**MANHEULLES**
*German Machine-Gun Blockhouse of concrete, in the middle of
the village, on left of the road.*

HATTONCHÂTEL CHURCH AND CLOISTER

## Hattonchâtel

Hattonchâtel stands on one of the promontories of the chain of hills which stretches from Verdun to Toul and which separates the Valley of the Meuse from the Plains of the Woëvre. It derives its name from a castle built in the 9th century by Hatton, Bishop of Verdun. The fortress has long been demolished. The church, erected as a collegiate in 1328, but united with the Collegiate Church of Apremont in 1707, remained standing until 1914. Since then it has been damaged by bombardments, especially the apse and north aisle. The little 15th century cloister, crossed by a public road, has suffered relatively little damage.

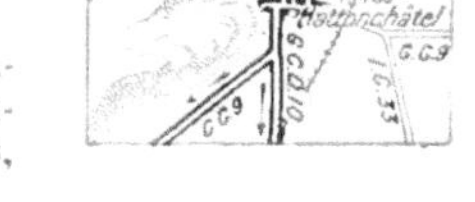

The church contained the tomb of G. de Haraucourt, Bishop of Verdun (16th century), and a remarkable altar piece. They were carried off by the Germans, but it is hoped that they will be returned.

The altar-screen, dating from 1523, is the earliest work attributed to

HATTONCHÂTEL CLOISTER

CELEBRATED ALTAR-SCREEN, BY LIGIER RICHIER, STOLEN BY THE
GERMANS FROM THE CHURCH OF HATTONCHÂTEL

Ligier Richier (*see p. 57*). It rested on a marble altar shaped like an antique
tomb.

This altar-screen, the projecting parts of which were of gold on a blue
background, is divided into three sections, separated by pilasters with finely
moulded bases.

On the central keystone, in the shape of a shield, are the arms of Duke
Antoine of Lorraine. Two medallions between the archivolt and the first
projection of the coping represent St. Peter and St. Paul.

The subjects of the three groups are: on the left, the **Carrying of the
Cross;** Christ, in a long flowing robe, is in the centre, while behind Him stands
Simon the Cyrenian wearing a pointed cap with turned-up edges ; around
stands a group of three women, two of whom are easily recognized—Mary
Magdalene with long hair falling over her shoulders, and Veronica holding the
Cloth of the Holy Face. Two executioners complete the scene.

In the centre of the altar-screen is **The Crucifixion.** In the foreground
is the swooning Virgin supported by St. John. Kneeling at the foot of the
Cross is Mary Magdalene, and opposite her, Stephaton holding the long reed
with a sponge dipped in vinegar. Lastly come the three soldiers of Pilate, one
of whom carries the spear which pierced Christ's side. On a pennant held by
the second soldier are inscribed the words which affirm the divinity of Christ:
"*Vere hic homo filius Dei erat.*"

The third section of the altar-screen represents the **Burial Scene.** In the
background is a bishop wearing a mitre, and kneeling at his feet a priest in a
surplice. According to custom, the sculptor has here represented the donor,
doubtless Gaucher or Gauthier Richeret, Dean of the Collegiate Church, whose

initials, " G. R.," frame the blazoned shield. The bishop is St. Maur, Bishop of Verdun, whose relics belonged to the Collegiate Church.

Unfortunate restorations were carried out in 1764 by Cellier Delatour, whose name appears on the background of the third picture. The date of the work (" A.D. 1.000.500.23.") is inscribed on each of the curtains of the four pilasters which surround it.

The 18th century pulpit is almost intact.

Behind the church there is a fine view over the Heights of the Meuse towards Apremont.

In the village square is the old guard-house with an arcade, and some old houses, most of which are uninhabitable.

At the end of the village, in the direction of Hattonville, are the ruins of

the old château. The cellars served as bomb-proof shelters, the walls being several yards thick.

From the terrace of the château is seen the immense Plain of Woëvre—partly occupied by the Germans from September, 1914.

## The Woëvre Plain

The Woëvre forms a district by itself, geologically rather than geographically, and corresponds approximately to the "*pays vabrensis*" of the Merovingians. It lies between the Heights of the Meuse and Moselle. The soil of marl and clay becomes a slough after rain, and numerous pools and hidden sheets of water, known locally as "*gorittis*," "*noues*," or "*crachettes*," make the ground slippery and treacherous.

Here may be followed step by step the stages of the Franco-American offensive of September, 1918 (*see pp.* 18–20), which reduced the whole salient of St. Mihiel, and advanced the lines several kilometres to the outworks of the Forts of Metz, thus placing the Allied forces in strong positions in readiness for the new offensive planned for November 16, which the signing of the

VIGNEULLES

*Entrance to concrete shelter near the Church, at the side of the road.*

Armistice on November 11 prevented from being carried out to overwhelming victory.

*After visiting Hattonhâtel, proceed to* **Vigneulles** *by a road which describes a large loop.*

*Leaving Hattonchâtel, a German cemetery will be seen on the left, beside the village cemetery.*

Vigneulles is a country town of considerable importance, built on the western fringe of the Woëvre Plain, at the foot of the chain of hills which separates the latter from the Valley of the Meuse. In the original plan of mobilisation it was to be the main French Headquarters.

Numerous houses have been destroyed.

*From Vigneulles to St. Mihiel there is a choice of two roads :* one, direct, via Chaillon (*Itinerary A, p.* 41) ; the other, less direct, passes through Apremont, Brûlé Wood and Ailly Wood, and is much more interesting (*Itinerary B, p.* 42).

## A.—From Vigneulles to St. Mihiel, via Chaillon

*At Vigneulles take G.C.D. 10 on the left, 100 yards from the church, and follow it for about 500 yards, then take G.C. 9 on the right, which passes through the village of Creüe.*

The woods which rise above the village form a kind of curtain, and the Germans, well aware of its importance (the Grand French Manœuvres of 1891 had taken place in this district), seized it at the end of September, 1914, and later built a light railway which formed their main line of communication with St. Mihiel. Hidden in this recess, the railway escaped observation and was worked, with but little damage, throughout the war.

*Leaving Creüe, the road follows the valley through which runs the Creüe Brook.*

*Before entering Chaillon, the tourist passes a German cemetery on the right.* Many of the houses in this village, as well as the church, were destroyed.

**GERMAN CEMETERY**

*This cemetery is in Mouton Wood, between Chaillon and St. Mihiel, on the left of the road when going towards St. Mihiel. One of the monuments represents a lion on a pedestal.*

*After Chaillon the road turns to the right and continues to follow the valley as far as the crossing of I.C. 62. It there climbs to the plateau, leaving the valley, which continues to the right in the direction of Spada.* This valley is the only one which crosses the Heights of the Meuse in their entire width, uniting the Plain of Woëvre with the river. It is the "Spada Pass," of immense strategical importance.

*At the top of the slope the wood is entered just beyond a military cemetery, 200 yards to the left of the road. Cross Mouton Wood, dotted with German graves, shelters, cantonments, etc.*

*Leaving the Varvinay road on the left, a 100 yards further on the tourist comes to a German cemetery by the roadside, with several monuments, one of which represents a lion on a large pedestal.*

*Follow the road as far as* **St. Mihiel.**

## B.—From Vigneulles to St. Mihiel, via Apremont, Brûlé Wood and Ailly Wood

*At Vigneulles, 100 yards from the church, take G.C.D. 10 to the left, in the direction of* Heudicourt.

*One kilometre from Vigneulles are* several large concrete shelters *to the right and left of the road.* Creüe Wood, *seen on the left of the road, 2 km. further on, is* full of German defence works.

The greatly damaged village of **Heudicourt** is next reached. Numerous houses were destroyed by fire. Beside the cemetery is a German cemetery. *On leaving the village there is a stone and concrete blockhouse.*

*Beyond Heudicourt, the road passes through* **Buxières** (ruined houses); **Buxèrulles** (slightly damaged), containing German cemetery; **Woinville** (German cemetery with a monument in the middle (*see photo below*), on the

APREMONT.   RUINS OF VILLAGE AND CHURCH

*right, before entering the village, and a roofless church)*; **Varnéville,** *entirely in ruins.   Leaving the village, the tourist passes several concrete shelters and blockhouses.*

*One and a half kilometres from Varnéville, G.C.D. 10, crosses G.C.D. 1 bis. Take the latter to the right towards Apremont.   800 yards from the fork, after crossing a bridge over the stream, the village of* **Apremont** *is reached.*

*Immediately after the bridge there is a very comfortable shelter of stone, cement and logs in a garden, behind a house on the right (the least damaged in the village).*

*Apremont is entirely in ruins.   Of the church, only a few broken walls remain.   In the Rue de l'Eglise, fifty yards from the church, to the left, near the*

APREMONT
*Shelter,
on the left
before crossing
stream, going
towards
Bouconville.*

*end of the village, is* a large concrete shelter, on the wall of which a German machine-gun has been carved (*see photo above*).

Apremont was a very important place during the war. At this point the road from St. Mihiel to Flirey and Pont-à-Mousson crosses the Vigneulles road, along which the tourist has just come, and which, beyond Apremont, goes down to Fort Gironville after skirting Fort Liouville on the right.

*G.C.D. 1 bis, turns to the right in the village, then mounts a steep slope towards* **Brûlé Wood.**

END OF APREMONT VILLAGE, GOING TOWARDS ST. MIHIEL
*On the right:* a Michelin sign; *in the background:* Hill and Fort of Gironville, Reine Forest and Vignot Wood.

*One kilometre beyond Apremont, in a quarry on the left of the road,* the Germans built a veritable village in concrete and cement, with deep shelters under the rocks. Terraces and flowering plants ornament the houses. The rooms are decorated with carved woodwork and tapestry. The furniture was either taken from the surrounding villages or made in rustic style (*see photo below*). *At the top of this camp, beyond the terrace of the* Officers' Mess, a cement staircase leads to a concrete trench which dominates the position in Brûlé Wood. The latter is furrowed with numerous German defence works.

ON THE ROAD TO ST. MIHIEL, 1 KM. FROM APREMONT

*German village built in the side of the quarries and occupied by the General commanding the sector. Above: Brûlé Wood.*

## Brûlé Wood

Lying almost on the edge of the Forest of Apremont, Brûlé Wood commanded the cross-roads on which the village of Apremont stands.

The German trenches were only fifty yards from the French lines at this point. For months, bombs, grenades and rockets made an inferno of the place. The proximity of the respective lines required the utmost precautions, constant

AMERICAN SOLDIERS IN GERMAN VILLAGE OF BRÛLÉ WOOD

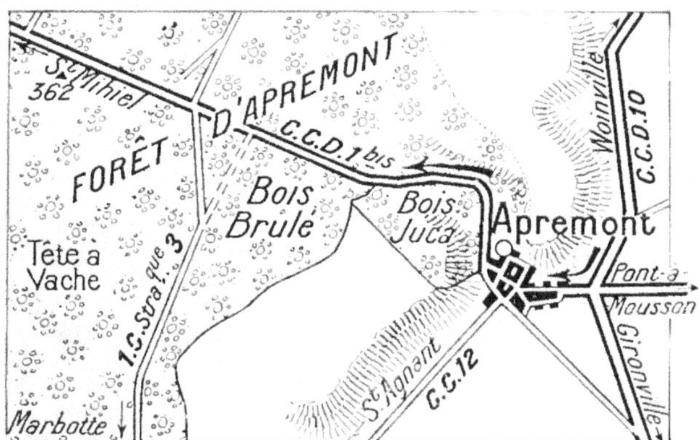


watching and listening, with finger on the trigger of the rifle, absolute silence, no sleep and no smoking (smoking might give an objective to the bombers). The nervous tension was so great that the average stay of a battalion was only eight days.

Brûlé Wood was the scene of the sublime rallying call "*Debout les Morts*" (Stand up, ye Dead!)—*see below*.

While early in April, 1915, important attacks were taking place in Ailly Wood, the 95th Infantry Regiment was ordered to create a diversion in Brûlé Wood. On April 5, 6 and 8 bloody fights took place for the possession of a trench. On the morning of the 8th the captured trench was consolidated, and the attacking troops relieved and sent in reserve to the second line.

Suddenly a strong German counter-attack was launched. The new occupants were thrown into confusion and, seized with panic, retreated through

GERMAN BLOCKHOUSE AT TÊTE-À-VACHE

FRENCH FIRST LINES AT TÊTE-À-VACHE

the trenches, when *Adjutant Jacques Péricard*, who had taken part in the action the day before but was now in reserve. called for volunteers from his company to face the enemy.  The trench was retaken after a prolonged and terrible struggle, in the course of which Péricard, feeling his men wavering and seeing only dead and wounded around, cried " *Debouts les Morts.*"

*Continue up the road.  Near the crest on the left, in a quarry, are* several concrete defence works which communicate with one another.

*The crossing of Strategic I.C. 3 is next reached.  Here leave the car and take the road to the left towards* **Marbotte.**  400 *yards further on the German*

GERMAN
BLOCKHOUSE
ON THE ROAD
TO
ST. MIHIEL,
*about
300 yds.
from Hill 362.*

first-line trenches, built entirely in concrete with numerous shelters and block-houses, *are reached. This is the crest of the* " **Tête-à-Vache** " position, which for so long formed a salient in the French lines. All the soldiers knew it because, when passing through the trenches on a level with this salient, it was necessary to stoop to avoid being seen by an observer at his loop-hole. Woe to the curious or the careless who risked walking upright past this point ! The ever-ready automatic spoke at once.

*A* 100 *yards beyond are the* French first-line trenches (equipments and soldiers' graves). All the ground here is torn up, and the woods are completely destroyed.

*Return to G.C.D.* 1 *bis, and follow it in the direction of St. Mihiel. All along the road are* numerous military works of all kinds, *especially across Ailly Wood.*

MILITARY KITCHEN IN AILLY WOOD, 1915

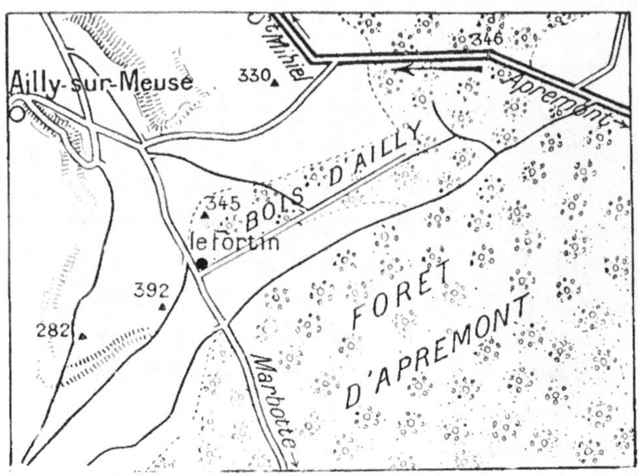



## Ailly Wood

Ailly Wood covers the brow of the hill, the southern slopes of which descend steeply towards a ravine.

Here the attacks took place which, between April 5 and 13, 1915, gave the French definite mastery of the position. The Germans held one corner of the wood and the outskirts at the foot of the slopes. The French trenches followed the ravine, mounted half-way up the unwooded part of the hill,

GERMAN TRENCH UNDER THE APREMONT-ST. MIHIEL ROAD

and ran alongside the wood. The entrenchment, known as the "*Le Fortin*," was in the corner. In the wood the German trenches rose in three tiers, linked together by narrow trenches. At certain points the Germans had constructed "*chevaux-de-frise*," twelve yards deep by two yards high.

The bombardment began on the morning of the 5th. The 75's opened breaches in the defences, and the observers, who were only 130 yards from the enemy line, gave accurate directions to the gunners. In their turn, 6-in. shells crushed the machine-gun emplacements, and at mid-day the explosion

IN AILLY WOOD

*German Post of Commandment at the side of the road, 4 km. from St. Mihiel.*

of five mine-fields annihilated the garrison and threw the enemy into a panic. A bayonet-attack was launched at once, without the firing of a single shot.

Two companies attacked on the western side of the wood, two others on the southern side.

The attack on the west was successful and, going beyond the third German line, reached the northern fringe of the wood. The machine-gunners, who followed the advance, at once took up their positions.

The attack on the south, after the first rush forward, was forced to withdraw slightly before an enfilading fire. At three o'clock in the afternoon the German artillery thundered ; at four o'clock a counter-attack was launched but failed; and at 5.30 the Germans tried to retake the lost ground by a terrific bombardment. In an hour and a half, on a front of 360 yards, twenty thousand shells of all sizes (4-in., 5.5-in., 6-in. and 8-in.) cut the French lines of communication, but failed to force a retreat. The attack was resumed next day, but in the evening, after fierce hand-to-hand fighting, the French still held the three lines of German trenches. On the 7th and 8th they repulsed eight counter-attacks, which left the shell-leveled ground in their hands.

On the 10th, after an artillery preparation lasting all day, a fresh attack

**IN AILLY WOOD SECTOR**

*German Defence-Works in quarry by roadside, 3 km. from St. Mihiel.*

was launched at seven o'clock in the evening. The whole of the wood was quickly occupied and immediately consolidated, in view of counter-attack. Five machine-guns, five trench-mortars, thousands of grenades and large quantities of equipment and stores, were left in the hands of the French.

From that time scarcely a month passed without some communiqué stating that the Germans had bombarded or counter-attacked Ailly Wood.

*St. Mihiel is entered via the Faubourg de Nancy, in which are the* burnt ruins of the Sénarmont Barracks.

*Follow the Rue Porte-à-Nancy, then the Rue Grande, as far as the Rue de l'Eglise, into which turn to the left to reach the Church of St. Etienne.*

## St. Mihiel during the War

On September 24, 1914, St. Mihiel was taken by the Germans, who held it until September 12, 1918.

Up to the latter date only one attempt was made to retake the town—the attack of November 17-20, 1914, during which a French unit succeeded in occupying the suburb of Chauvoncourt, but was forced to retire as the Germans had mined this section.

The Franco-American offensive of September, 1918, finally cleared St. Mihiel.

General Pershing, in the disposition of his forces, generously arranged that

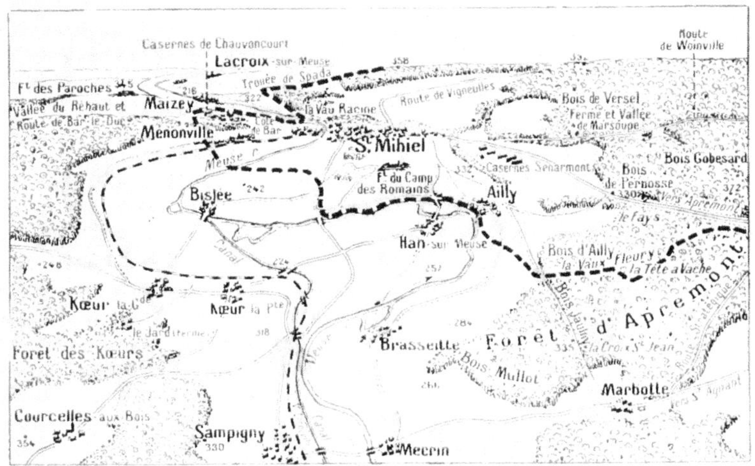


PANORAMIC SKETCH OF THE ST. MIHIEL REGION, SHOWING THE FRONT-LINE UNTIL SEPT. 12, 1918

a French regiment, the 25th Colonial, should have the honour of being the first to enter St. Mihiel. The Prime Minister's son, Captain Michel Clémenceau, was among those who marched into the town.

On the whole the town had suffered little. The bridges had been blown up, trenches cut up the streets, and a German narrow-gauge railway ran through the town. The monument of 1870, "*Aux Morts pour la Patrie*," was damaged. As everywhere else, all copper had been removed, the machinery had disappeared or had been broken, while the optical-glass factory and the copper foundry had ceased to exist.

On Friday, September 13, General Pershing, accompanied by General Pétain and Mr. Baker, American Secretary of State for War, visited St. Mihiel. The next day President Poincaré, in his turn, paid homage to the valiant city.

Little by little, when the first excitement was over, the inhabitants told the story of the occupation; of the war levies imposed by the Germans, as in every town which they had occupied: first a million francs in 1914, when the commandeering without payment or vouchers; the fines (20 francs for omitting to salute an officer); children forced to work in the trenches; people sent to prison, and even to the convict prison on the slightest pretext; an abbé deported as a hostage because he had said in a sermon, "*After the thorns will come the roses*:" a whole family placed in solitary confinement for forty days because they were suspected of having telephoned to the French, etc.,

**ST. MIHIEL DELIVERED**

*Group of children in French Officers' Car on Sept. 13, 1918.*

not to mention the systematic looting and removal of objects of art, pictures and silver.

On Tuesday, the 10th, the Germans, knowing the attack was imminent, made their final preparations for departure. On the 11th they ordered the inhabitants, on pain of death, to remain indoors until noon on the following day.

During the night of the 11th they blew up the bridges and removed their guns. On the morning of the 12th the French entered the town.

Several days later the American Headquarters which, until then, had been at Souilly, on the road from Verdun to Bar-le-Duc, moved into St. Mihiel.

# A VISIT TO ST. MIHIEL

## ITINERARY

*Enter the town via the* Faubourg de Nancy *and the* Rue Porte-à-Nancy (1).
St. Etienne Church (A) ; Place Ligier-Richier (4) ; Hôtel-de-Ville (H) ; Rue
Porte-à-Metz (6) ; Promenade des Capucins ; Les Sept-Roches.

*Cross the Meuse by the* temporary bridge (14), Chauvoncourt, *and* Paroches
Fort.

*Return to St. Mihiel by the temporary bridge.* Place des Halles (17) ; Place
du Collège (20) ; Church of St. Michel (B) ; *and* Hôtel de la Division.

*Leave St. Mihiel by the Commercy road.*

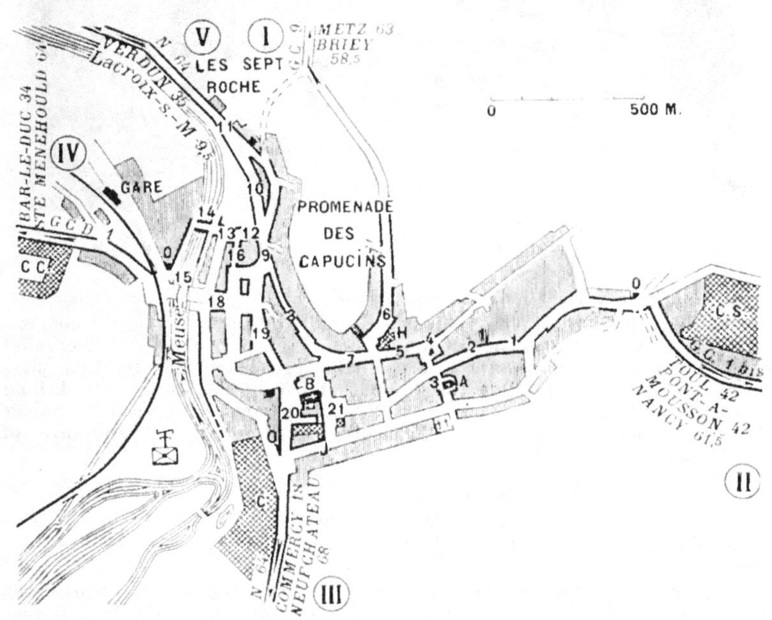


THE STREETS TO BE FOLLOWED ARE SHOWN BY THICK BLACK LINES

### Plan of St. Mihiel

*Arbitrary Signs*

| | | | |
|---|---|---|---|
| A. | St. Etienne Church. | 7. | Rue Carnot. |
| B. | St. Michel Church. | 8. | Rue du Général Blaise. |
| C. | Cavalry Barracks. | 9. | Rue Haute des Fosses. |
| CC. | Chauvoncourt Barracks. | 10. | Rue des Annonciades. |
| CS. | Sénarmont Barracks. | 11. | Avenue des Roches. |
| H. | Hôtel-de-Ville. | 12. | Place du Marché. |
| J. | Palais de Justice. | 13. | Place du Manege. |
| O. | Octrois. | 14. | Temporary Bridge. |
| | | 15. | Destroyed Bridge. |
| 1. | Rue Porte-à-Nancy. | 16. | Rue du Sauley. |
| 2. | Rue Grande. | 17. | Place des Halles. |
| 3. | Rue de l'Eglise. | 18. | Rue du Pont. |
| 4. | Place Ligier-Richier. | 19. | Rue Notre-Dame. |
| 5. | Rue de la Vaux. | 20. | Place du Collège. |
| 6. | Rue Porte-à-Metz. | 21. | Place aux Moines. |

THE "SEPULCHRE," *by Ligier Richier*, IN ST. ETIENNE'S CHURCH

Starting point: The Church of St. Etienne.

The Church of St. Etienne, often called the "Eglise du Bourg," contains several remarkable Renaissance sculptures, chief among which are a **bas-relief** in St. Joseph's Chapel (*photo, p.* 58), a large **reredos** behind the high altar (*photo, p.* 58), and above all, in the central bay of the south aisle, behind a railing (*photo, p.* 57) in a sort of grotto or crypt, the *chef-d'oeuvre* of Ligier Richier, commonly known by the incorrect title of the **"Sepulchre of St. Mihiel."**

## The "Sepulchre," by Ligier Richier

This comprises a group of thirteen figures, rather more than life size, executed between 1554 and 1564, and placed after Richier's death in the church, where it stands to-day (*photo above*).

The figures are arranged as follows : *on the left*, Salome lays in the coffin, the shroud which is to enwrap Christ, while two disciples, Joseph and Nicodemus, carrying the body of their Divine Master, stand in the foreground. Nicodemus carries the body by the shoulders, while the unsupported head rests against his arm. Joseph of Arimathæa, one knee on the ground, supports the legs of Our Lord. Near him Mary Magdalene helps to carry the feet, which she touches with her lips. In the background the Virgin, leaning on St. John, and Mary, the wife of Cleophas, turn a last look on Christ.

Between Mary, the wife of Cleophas, and Salome, stands an angel bearing the Cross—an unfounded tradition says that this is a portrait of the artist.

*On the right of the central group, and next to Nicodemus, a woman*, often called Veronica, carries the crown of thorns. Behind her, in the background, two men-at-arms are casting lots for the seamless coat. *At the other end, on the right*, a Roman officer, often, without justification, called the Centurion, is seated on a shield with a sword in his left hand. He is the captain of the guard in charge of the tomb.

This is a strong, touching work by a French master-sculptor, who had not yet come under Italian influence and methods.

THE RAILING OF LIGIER RICHIER'S " SEPULCHRE "

## Ligier Richier

Numerous legends surround the life of the " Master of St. Mihiel." The only son of Jean Richier, a master-sculptor, he was born at St. Mihiel about 1500. Brought up as a Catholic, he was converted to Calvinism about 1560. There is a legend that Michael Angelo came to St. Mihiel, admired the work of the boy Ligier Richier, and took him to Rome ; but it is known that Michael Angelo never visited Lorraine.

Ligier Richier, not being able to carry out his commissions single-handed, gathered around him apprentices and companions, who have been called his brothers. It is true that he had a son (Gerald) in 1534, and that the latter worked in his father's studio, and had in his turn five sons, also sculptors, who settled in Nancy, Metz, Lyons and Grenoble. In 1764, in consequence of the persecution of the Protestants, he settled in Geneva, where he died about 1567.

Numerous groups of sacred figures, scattered over this district, attest the happy skill of Ligier Richier : a **reredos** of many-coloured stone in the church at Hattonchâtel (p. 38) ; **Christ Crucified** between the Virgin and St. John, in the church at Genicourt, on the road from Verdun to St. Mihiel : **Group of Notre Dame-de-Pitié,** in the Sacré-Cœur Chapel of the church at Étain ; a

**Calvary** (six statues of wood variously coloured) in the chapel of the new cemetery at Briey ; a large **Christ carrying the Cross** in the Chapel of Notre-Dame-de-Pitié in the Church of St. Laurent, at Pont-á-Mousson (p. 92) ; and lastly, in St. Mihiel itself, two of his masterpieces : **"The Swooning Virgin,"** in the church of St. Mihiel (p. 67) and the important group, **"The Sepulchre"** in the Church of St. Etienne (p. 56).

ST. MIHIEL.    LIGIER-RICHIER SQUARE

*After visiting the Church of St. Etienne take on the right the Rue de l'Eglise, at No. 3, of which there is a curious old house.*

*Follow the continuation of the Rue de l'Eglise (Rue du Général Audéoud), which leads to the* **Place Ligier-Richier.**

*The statue of Ligier Richier used to stand in the centre of this Place. It was removed by the Germans during the occupation.*

*At the left-hand corner of the Place take the Rue de la Vaux. On the right of the street is the* **17th century Hôtel-de-Ville,** *at the corner of the Rue Porte-à-Metz. On the left of the Rue Porte-à-Metz, stairs lead to the public*

ST. MIHIEL TOWN HALL

MONUMENT IN GERMAN CEMETERY (*see below*)

garden called "La Promenade des Capucins," which overlooks the town and the valley of the Meuse (*fine panorama*).

*If the tourist enters St. Mihiel by the Rue Porte-à-Metz, he will see a large* cemetery containing more than two thousand granite monuments, opposite the first houses of the town. Over six thousand Germans were buried there, killed for the most part during 1915, and a few in 1916.

No Frenchman, soldier or civilian, has been buried in this cemetery.

*The cemetery can also be reached by returning along the Rue Porte-à-Metz as far as the last houses.*

GERMAN CEMETERY AT THE PORTE-À-METZ ENTRANCE TO ST. MIHIEL (*see above*)

THE "MAISON AUX BOEUFS"
(No. 3, Rue de la Vaux.)

*Return to the Rue de la Vaux, which take on the right.*

*At No. 3, on the right, is an old Renaissance house with curious gargoyles ; and at No. 2, opposite, a curious old house.*

*At the beginning of Rue Carnot, which is a continuation of Rue de la Vaux, see the 16th century house called "Du Narrateur," at No. 36.*

*Opposite this house, take the terraced Rue du Général Blaise (old house at No. 30).*

*Follow its continuation (Rue Haute des Fosses), at No. 7 of which is Ligier Richier's house.*

*Take the Rue des Annonciades as far as the Avenue des Roches, which leads to the "Seven Rocks," situated at the gates of the town (see p. 62).*

RENAISSANCE HOUSE
(No. 36, Rue Carnot.)

## The Seven Rocks

On the edge of the Verdun road, which is an extension of the Avenue des Roches, and overhanging the Meuse just beyond the town, rise seven rocks, known as the " Cliffs of St. Mihiel."

In the first, surmounted by a stone cross, a grotto has been hollowed out, containing a tomb in which lies a great stone Christ. A staircase in the rock gives easy access to it.

THE COAT OF ARMS OF ST. MIHIEL

*Three rocks argent on an azure field, two in chief, and one in pile.*

This calvary is a place of pilgrimage on Good Friday. The sixth rock, worn away by water, resembles a gigantic mushroom and is called the " Devil's Table."

These rocks are represented in the St. Mihiel coat of arms, which is : " Three rocks argent on an azure field, two in chief and one in pile."

Across the intervening wooded slopes are seen the large buildings of the old Benedictine Abbey of St. Michel, of which the name St. Mihiel is a corruption.

Founded in 709 on a site known to-day as St. Christopher's Farm, in Vieux-Montier Wood, it was transferred in 819 to the village of Godinécourt, which then took the name of St. Michel or St. Mihiel. It was closed in 1791.

A romanesque tower, dating from about 1060, dominates the abbey.

DESTROYED BRIDGE OVER THE MEUSE

*Return by the Avenue des Roches to the market ; behind the latter*, take the Place du Manège, which leads to the temporary bridge across the Meuse.

Cross the bridge and take G.C.D. 1, which passes through the suburb of Chauvoncourt.

## Chauvoncourt

Chauvoncourt, occupied by the Germans from the beginning of their advance in September, 1914, was an important bridgehead which the French had an interest in retaking. Its capture and subsequent evacuation (November 16–18) are famous.

In the evening of November 16, French heavy batteries took up their position at Fresnes-au-Mont, on the left bank of the Meuse, five miles from St. Mihiel; but before attacking, the German howitzers on the Paroches position had been destroyed.

The bombardment began at dead of night. Four hundred shells fell on the enemy, causing the Bavarian ammunition dump to explode. At dawn, French infantry, massed in the peninsula of Les Romains, crossed the Meuse on a pontoon bridge, whilst the cavalry on the Fresnes road threatened Chauvoncourt from the west. By ten o'clock the infantry were in sight of the village. The Bavarians advanced by successive rushes—at each of which they fired a salvo—then halted behind a little glen. The fight became a fusilade, and would have continued indefinitely but for the arrival of the French dragoons, who, with lances fixed, charged furiously. The enemy, afraid of being cut off, retreated, followed by the cavalry, who began the siege of the houses. Every window, door and roof sheltered a Bavarian marksman. All day on the 17th the fighting continued in favour of the French, who by night occupied the western part of Chauvoncourt and slept in a French barracks. On the left bank of the Meuse the Germans, two hundred of whom had surrendered, now occupied only a few ruined houses.

But at five o'clock on the morning of the 18th an explosion was heard. At the end of the main street three houses, luckily unoccupied, had been wrecked. Orders were at once given to evacuate the occupied portion of the town, which proved to be a wise precaution, for at eight o'clock the whole southwest portion blew up, over an area of four acres. No soldiers were killed, but civilians, who stayed on in spite of orders to the contrary, were victims of their own imprudence.

Trenches and shelters are to be seen all along the road.

RUINS OF CHAUVONCOURT BARRACKS

PANORAMIC VIEW OF ST. MIHIEL AND THE VALLEY OF THE MEUSE

PAROCHES VILLAGE CHURCH

*Leaving Chauvoncourt, take to the right G.C. 34, which leads to* **Paroches.**

The village of **Paroches** is an absolute ruin. The 14th century church, with the exception of part of the belfry, has been almost entirely destroyed (*photo opposite*).

*At the end of the village, on the right, near the Calvary and skirting the wall of the last house, there is a military cemetery. Take, on the*

SEEN FROM THE TOP OF PAROCHES FORT

left, *the narrow road to* **Fort Paroches.** Around the fort are numerous defence works and the graves of French soldiers.

**Paroches Fort,** built to protect the approaches to Spada Pass, which Troyon Fort defended on the north, is an old masonry fort. Visit the shelters, inner works in concrete.

From the summit there is a fine sweeping view (*panorama above*) over the

ENTRANCE TO PAROCHES FORT

THE PLACE DES HALLES, ST. MIHIEL

battlefield, the Valley of the Meuse, St. Mihiel, the "Camp des Romains" Fort (p. 65), Versel Wood and Spada Pass.

*Return to St. Mihiel by the same road as far as the temporary bridge over the Meuse and the Place du Manège.*

*From the Place du Manège, take, on the right, the Rue du Sauley, which leads to the Place des Halles.*

*To the right, on reaching the "Place," is the Rue du Pont, which leads to the ruined bridge. Turn to the left, cross the "Place," and take the Rue Notre-Dame on the right.*

*At No. 1 of this street is a 15th century house with a polygonal turret. Opposite at No. 2 is a 11th-15th century house, known locally as the "Maison du Roi."*

THE MAISON DU ROI

THE PLACE DU COLLÈGE AND CHURCH OF ST. MICHEL

*Follow the Rue des Carmes, which is a continuation of the Rue Notre-Dame, then the Place des Regrets to the Place du Collège.*

*On the left stands the* **Church of St. Michel.**

This 17th century church is recognisable by its old Romanesque tower, which forms a vestibule in front of the building. It has three naves with side-chapels divided into five bays by fluted columns.

Note the **fine organ** (*photo. p. 68*), the pipes of which were removed by the Germans. In the Chapel of the Baptismal Fonts is a stone Cupid holding skulls.

The Church of St. Michel contains one of the finest of Ligier Richier's works,

THE SACRISTY, CHURCH OF ST. MICHEL
*As the Germans left it.*

THE ORGAN,
CHURCH OF
ST. MICHEL

known as "*The Swoon*," or "*The Fainting Virgin*." It stands to the right of the choir in a chapel with a door.

In the Middle Ages it was customary to represent the Virgin standing in contemplation of the wounds of her Divine Son, as described in the famous chant, *Stabat Mater Dolorosa*. In the 15th century, on the contrary, the Virgin was generally represented as described in the Gospel of Nicodemus. In Richier's group we see the Virgin, supported by St. John, fainting at the foot of the Cross. The extreme simplicity of the work renders it most pathetic.

This work is only a fragment of a much larger group which comprised a large Crucifix, and on either side of the Virgin, St. Longin, Mary Magdalene and

THE HÔTEL DE LA DIVISION AND CHEVET OF
ST. MICHEL CHURCH

" CAMP-DES-ROMAINS " FORT

four angels, each holding a chalice to catch the Saviour's blood. The work was in painted walnut, as had been the custom from the Middle Ages, but the worm-eaten wood gradually crumbled away. In 1720 the Benedictines managed to save the crucifix and the group of "The Swoon." The crucifix is supposed to have been burnt during the Revolution (1792). Now all that remains is a moulding of Christ's head.

*Leaving the Church, skirt the front of the adjoining Collège, and pass under the arch of the Palais-de-Justice, thus reaching the* **Place des Moines.** In this square is the fine façade of the old abbey, restored in the 17th and 18th centuries, the buildings of which have been transformed into the Hôtel de la Division, Palais-de-Justice and prison. Above the latter is the famous monastic library containing 13,000 volumes and valuable manuscripts.

*Besides the Hôtel de la Division is the chevet of St. Michel Church, looking on to the square of that name. By turning to the left in the latter, the tourist comes back to the Place du Collège, which cross to take the Commercy road (N. 64).*

*On leaving St. Mihiel, N. 64 climbs up a steep slope. A mile from the town, on the left, is a concrete blockhouse at the corner of the Commercy road and that leading to* Fort Camp-des-Romains. *Take the latter to the fort.*

### Fort of the Camp-des-Romains

This is one of the two forts which protect St. Mihiel. Standing on the end of a narrow peninsula formed by a loop in the Meuse, it dominates the town from a height of 450 feet above the valley. (The hill itself is 1,200 feet high.) It owes its name to the remains of Roman entrenchments, still existing when the fort was built.

When the German Army of Metz occupied St. Mihiel on September 24, 1914, and crossed the Meuse, the Fort of the Camp des Romains remained isolated, without troops in the plain to defend it, and absolutely dependent on its own guns. The Germans left it alone for the time being, confident of being able to take it whenever they wished. The 16th Corps hastened to the rescue, but

stopped in front of St. Mihiel. The Germans finally dug themselves in and were able, from a position near the town, to begin the bombardment of the fort with the aid of Austrian heavy guns.

The guns were very quickly placed in position, and in a few days they silenced those of the French fort, the turrets and bastions of which were destroyed. In the end the heroic garrison were smoked out by the enemy, who had reached the base of the fort. When the surviving defenders, half suffocated, were able to leave the ruins, the Germans presented arms as a tribute of admiration for their valour, and permitted the captured officers to retain their swords.

All this sector, with a few slight changes, was to remain in the hands of the Germans until September, 1918.

In spite of the terrific gunfire to which it was subjected, the fort was not completely destroyed. In the moats and on the bastions are numerous concrete blockhouses built by the Germans. Near the entrance is the grave of Captain of Artillery Cordebard, killed in 1914.

From the fort there is a fine view on all sides over the valley of the Meuse and the Forest of Apremont.

*Return to N. 64 which descends in a long zig-zag to the Meuse, which it crosses.*

*The road passes through* Sampigny—considerably battered—where President Poincaré's country house was completely ruined by the German bombardments.

*It next crosses the railway before entering Vadonville and again on leaving* that village.

Lerouville, then Commercy, *are soon reached.* The night should be spent at the latter *(see information on the fly-leaf inside cover).*

Commercy is of no particular interest from a picturesque or artistic point of view, but its " Madeleine " cakes enjoy a world-wide reputation.

PRESIDENT POINCAIRÉ'S HOUSE AT SAMPIGNY

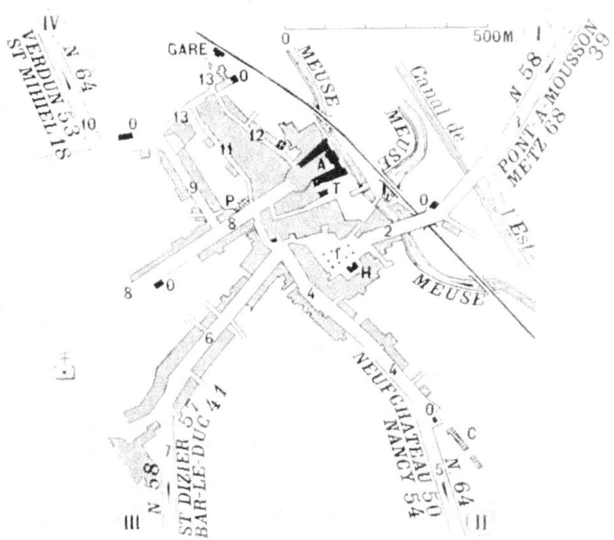


## Plan of Commercy

### *Arbitrary Signs*

A. —Old Château, now a Barracks.
C.— Barracks.
H.—Hôtel-de-Ville.
O.—Octrois.
P.—Sous-Préfecture.
T.—Theatre.

1.— Place de l'Hôtel-de-Ville.
2.— Rue du Pont des Religieuses.
3.—Road to Metz.

4. Rue des Capucins.
5.—Road to Nancy.
6.— Rue Levée-de-Breuil.
7. Road to Bar-le-Duc.
8.— Rue Carnot.
9. Rue de Lisle.
10. Road to Verdun.
11. Rue du Four.
12.— Rue de l'Église.
13. Rue de la Gare.

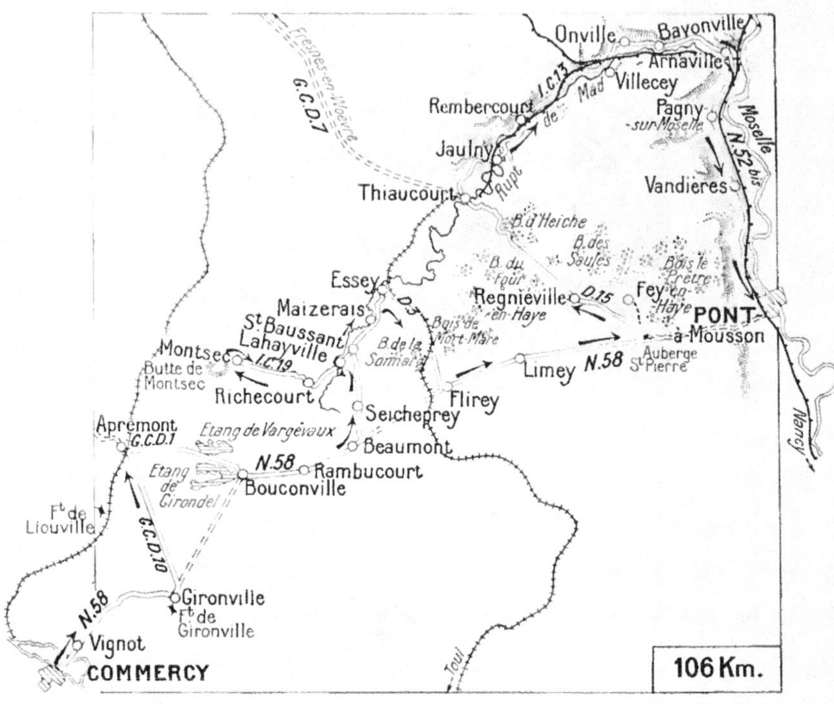


## SECOND DAY

## COMMERCY—PONT-À-MOUSSON—METZ

### A.—Commercy to Pont-à-Mousson (*See above.*)

### B.—Pont-à-Mousson to Metz (*See p.* 109.)

*Leave Commercy (Place de l'Hôtel de Ville) by the Rue du Pont-des-Religieuses which, after crossing the Meuse, joins N. 58. Take the latter.*

*Pass through* **Vignot** *(2 km.) and enter* **Vignot Wood** *(French gun emplacements).*

*After crossing the wood the tourist approaches* **Gironville.** *Before entering the village, on the crest to the right, is* **Gironville Fort.**

*In the village immediately beyond the church, N. 58 turns to the left. Continue along it for 500 yards beyond the church, then leave it where it turns to the right, and take G.C.D. 10 towards* **Apremont.**

*The road crosses the old French and German lines (shelters and block-*

BOUCONVILLE VILLAGE

houses), *then rejoins, 200 yards east of Apremont, G.D.C. 1 bis, which take to the right towards* **Bouconville.**

*Before going to Bouconville, visit Apremont, Brûlé Wood and Ailly Wood, if this was not done on the first day (see pp. 42-52).*

*Notice, in succession, on the left, the ruins of* **Loupmont Village,** 1,500 *yards from the road ;* **Montsec,** *further north, dominating the whole district ;* and **Vargévaux Pond,** *near the road.*

*Follow G.C.D. 1 bis, to* **Bouconville.** *Enter the village, leaving the fine* **Girondel Pond** *on the right.*

NO-MAN'S LAND, NEAR RAMBUCOURT

BEAUMONT CHURCH

The 13th-14th century Church of Bouconville with its three naves is very curious.

In the cemetery are numerous French graves.

The front line, after passing south of Apremont, continued first to the right of the Bouconville road, then crossed the road to the west of Vargévaux Pond, making a bend to include the latter within the French lines, as also the village of Xivray, which was the junction of the French armies with the American divisions. It then passed through the hamlet of Seicheprey, and at Flirey rejoined the main road leading to Pont-à-Mousson.

*Keep along the road towards* **Rambucourt.**

*A little before this village, N. 58 is picked up again.* All the way the road is camouflaged and bordered by numerous trenches. **Rambucourt** was badly damaged. Numerous shelters were made along the road against the houses, the basements of which were occupied.

*After passing through Rambucourt, N. 58 leads to* **Beaumont** (in ruins).

AMERICAN AMMUNITION CONVOY ENTERING SEICHEPREY

ROMANESQUE CHURCH OF SEICHEPREY

Notice the shelters in the houses. The curious church suffered badly (*photo, p. 74*).

*Four hundred yards beyond Beaumont, leave the National road and take the Seicheprey road on the left.*

Trenches, shelters and gun-emplacements are met with, especially in a hollow on the left. **Seicheprey** *is next reached.*

This village was the scene of one of the first successes of the American Army. The Germans had taken it by surprise in April, 1918, and had kept it for some time, when it was retaken by the 26th (New England) Division.

Part of the belfry of the 12th century church is still standing (*photo above*).

*Near the church the road bends to the right and goes towards* **St. Baussant.** *Half a mile further on* the French and German first line trenches *are crossed. On entering St. Baussant,* notice in a house on the right in front of the stream

SEICHEPREY.   THE MAIN STREET ON SEPT. 12, 1918

ST. BAUSSANT VILLAGE, ON ENTERING

*The house on the left was transformed into a concrete Blockhouse.*
*Above the ruined wall : Loopholes for the machine-guns.*
*In the background : Ruins of old Castle.*

a large machine-gun blockhouse in concrete. The loopholes are on a level with the roof.

**St. Baussant** is almost entirely in ruins. *To the right, on the hill, stood the old château,* of which only a few broken walls are left.

This village, being an important road junction, had been strongly fortified by the Germans. It is one of the places where the American tanks performed wonders, taking the position in less than half an hour.

The last house of the village, at the *I.C.* 13 crossing, on the right, bears the inscription " *Café Hocquard.*" Here are three large concrete shelters, the walls of which are five feet thick. *At the fork in the road is* a machine-gun blockhouse in concrete.

ST. BAUSSANT VILLAGE AND RUINS OF THE OLD CASTLE

RICHECOURT VILLAGE AND THE RUPT-DE-MAD STREAM
*In the foreground : The bridge over which I.C. 19 passes.*
*In the background : Montsec Hill.*

*Take I.C. 13 to the left, which follows the Rupt-de-Mad stream and becomes G.C. 33 on reaching* **Lahayville.** *1,600 yards beyond this village (greatly damaged), leave G.C. 33 and take I.C. 19 to the right. After crossing the bridge over the Rupt-de-Mad,* **Richecourt,** *razed to the ground, is reached (photo above).*

*Near a house on the left, at the end of the village, is a German concrete shelter with the inscription " Pommernburg." Other shelters of less importance are to be seen among the ruins.*

GERMAN SHELTER IN RICHECOURT VILLAGE
*This concrete shelter is seen on leaving the village by the road to Montsec. Over the door is the word " Pommernburg." The village is a heap of ruins.*

MONTSEC
*German Telephone Exchange on the road to Woinville.*

*Follow I.C. 19 as far as* **Montsec** *(3 km.).*

The village of Montsec is at the foot of the hill ; it was badly damaged.

**Montsec Ridge,** or Hill 380, made a first-class observation-post for the Germans, as it dominates the whole district from Apremont to Flirey.

Montsec was the scene of the fiercest fighting on June 17, 1916. The French were unable to take it on account of its formidable defences. From that time no surprise-attacks took place in this district.

On the crest, the Germans had constructed a system of tunnels, the entries of which overlooked the region of Heudicourt—Buxières, and at the end of which chimneys over 30 feet in height opened in the summit of the hill.

MONTSEC   RUINS OF CHURCH AND VILLAGE

MONTSEC
*German Signalling Post.*

The observers climbed up the chimneys by means of ladders and directed the firing of the artillery, which was massed in the surrounding woods.

The system of trenches and shelters was remarkable. In some places the shelters were furnished with electric light.

*To visit the military works of Montsec, go beyond the village along I.C. 19 and stop at the last houses on the left, where there is an enormous concrete shelter, which served as an artillery telephone exchange, (photo, p. 78). A narrow road leads from this shelter to the entrances of the tunnels on the crest. (Time required to visit : one hour.)*

AMERICAN COLUMNS MARCHING TOWARDS MONTSEC (SEPT. 13, 1918)

*After visiting Montsec return to St. Baussant by the same roads (I.C. 19 to Richecourt, then G.C. 33 and I.C. 13).*

*Follow I.C. 13 beyond St. Baussant to* **Maizerais** *(completely ruined), seen on the left of the road, and* **Essey** *on the Rupt-de-Mad stream.*

From December, 1916, the village of Essey was close to the front, and occupied by the Germans. The inhabitants and mayor remained during the occupation, but were forbidden, on pain of death, to go more than a short distance from their homes.

*In the village, at the corner of the Rue Béquille and the Grand Rue, is a concrete blockhouse.*

The church was partly destroyed. On its north front, protected by the church belfry, were the Kommandantur's quarters—an important concrete construction with walls five feet thick.

*After visiting Essey, take D. 3, which passes in front of the church, and follow it towards* **Flirey.**

SONNARD WOOD

*American Cemetery. In the background, at the foot of the larger trees, is the road D. 3.*

*To the right and left are* numerous shelters. *Turn to the right alongside Sonnard Wood, beside which, 50 yards from the road, are* an American cemetery *and, on the left, Mort-Mare Wood.*

**Mort-Mare Wood** is famous for the terrible struggles that took place for its possession.

It was while reconnoitring over this wood in an aeroplane that *Senator Reymond* was killed on October 22, 1914. He was returning from a flight over Mars-la-Tour, Chambley and Thiaucourt, with Pilot Adjutant Clamadieu, and the machine was turning to the right of the southern edge of the wood, when it was seen descending, apparently normally, between the French and German lines. Machine-guns at once opened fire ; the Adjutant was killed and Senator Reymond wounded. The French came out of their trenches and a fierce struggle, which lasted until night, took place round the machine. Only then was Reymond able to crawl into the lines, while the French carried back the body of the Adjutant.

Reymond was taken to the hospital at Toul, and was able, before he died, to give an exact account of the mission in the fulfilment of which he had met such a glorious end.

*On reaching the crest, the road crosses the* old German first-lines (concrete blockhouses). **Flirey** *next comes into sight.*

This village, which formed part of the first French lines from 1914, is almost completely in ruins, while the whole country around is laid waste.

*On the right are seen* the ruins of Toul-Thiaucourt railway bridge.

*Half a mile from the village, keep along D. 3, to visit the* famous **Flirey Quarry,** where there are numerous shelters and French graves. The surrounding woods contain the emplacements of several batteries.

*Return to Flirey and take, on the right, N. 58 towards* **Pont-à-Mousson.**

RUINS OF FLIREY VILLAGE
*N. 58, seen in the photo, passes through the village.*

FLIREY QUARRY

*In the background : D. 3 and Sonnard Wood.*

*One kilometre from Flirey, at the top of the crest, on the right, is a* military cemetery. *The road runs parallel with the* old French first-lines, which followed the crest on the left.

*At the entrance to* **Limey,** *through which* N. 58 *runs, there is* a large French cemetery *on the left.*

The village of Limey, famous for the hard and bloody battles fought there in September, 1914, is in ruins. The west front of the church was torn open. Numerous shelters are seen, two of them, in cement, being very large ; the first, *in the middle of the village on the right of the main road ;* the other, a machine-gun blockhouse, *in the last house on the right.*

*Beyond Limey the road crosses a vast wooded district, known as* **La Haye,** which covers the whole plateau.

*Two and a half kilometres from Limey, on the left side of the road,* a place called **Fond-des-Vaux** contains numerous French shelters (several in concrete), and also a military cemetery. This is **Lampe Camp.**

LIMEY VILLAGE

*Concrete shelter on the right of N. 58, when coming from Flirey
in the middle of the village.*

LAMPE CAMP
*At the back, on the left : N. 58.*

Three hundred yards further along the road there are American graves *to the right.*

*The* **Inn of St. Pierre** *is next reached, from which D. 15 leaves to the left towards* **Thiaucourt.** *Take this road.*

Throughout the war St. Pierre Inn, which is at the entrance to **Prêtre Wood,** was the nearest dressing-station to the front. The buildings suffered

ST. PIERRE INN DRESSING STATION
*Arrival of wounded soldier.*

little, thanks to the sheltering forest. *Prêtre Wood will be visited on leaving Pont-à-Mousson.*

*Eight hundred yards from the inn, to the right of D. 15, the* **Fey-en-Haye** *road debouches. This road is not available for motors. A visit to the village is interesting, as it was in the first French lines (distance there and back, 3 km.).*

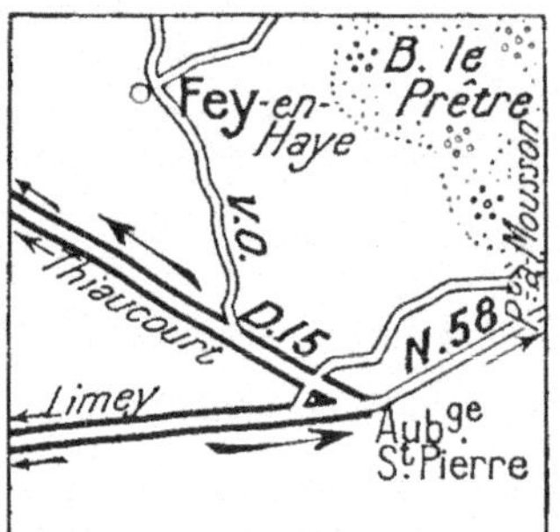



**Fey-en-Haye** *is about* 100 *yards from the western edge of* **Prêtre Wood.** At the end of September, 1914, a bloody engagement took place there. Up to the end of March, 1915, this unfortunate village was continuously bombarded, and it was entirely demolished when, on April 2, 1915, it was taken by a French battalion (169th Infantry). Its capture was the prelude to the last series of attacks which, after seven months of terrific fighting, ended on May 31, in the capture of **Prêtre Wood.**

Fey-en-Haye is now merely a heap of ruins. A number of trenches run through it, and a few shelters still exist.

*After coming back to D. 15, continue along it as far as* **Regniéville,** a village of which nothing remains but part of the belfry of the church (*photo, p. 85*).

RECNIÉVILLE VILLAGE AND THE RUINED BELFRY OF THE CHURCH

For a long time Regniéville was the advance-post of the French line between Mort-Mare Wood, *on the left*, and Prêtre Wood, *on the right*. At the beginning of April, 1915, the French advance was especially dangerous for the enemy, whose counter-attacks became more frequent. It was evident that the slightest advance in the direction of Thiaucourt would hamper the German communications between Metz and St. Mihiel, and would hinder the revictualling of the troops as well as the steady supply of reinforcements and munitions. That is why, on April 9, the Germans made fifteen successive attacks to drive the French from their trenches and the edge of Mort-Mare Wood.

*Keep on towards* **Thiaucourt.** *To the left of D. 15 there is an* American cemetery (*photo below*), 500 *yards from Regniéville.* 1 *km. further on, fifty*

AMERICAN GRAVES AT REGNIÉVILLE

*Smashed Renault Tank at the edge of the wood.*

*yards from the road, and before entering* **Four Wood**, *lies* a derelict Renault tank *(photo above), and beside it* the graves of its drivers. *In the Wood are* numerous German gun emplacements.

*On reaching the crest (Hill 340), on the right, alongside Saules Wood, are* two German gun shelters.

*Further on, at the "milestone" 4 km. from Thiaucourt, is a concrete block-house (photo below).*

*From I.C. 13, D. 15 descends in a large bend across* **Heiche Wood**.

*One kilometre from Thiaucourt, in a ravine on the right of D. 15, there stood* a large railway station and an important German military depot. *On the other side of the road there is a* German cemetery containing 600 graves.

**Thiaucourt,** altitude 750 feet, stands in an amphitheatre, in the centre of a loop described by the Rupt-de-Mad stream.

A large number of its houses are in ruins, especially on the banks of the Rupt.

Thiaucourt was a rest-camp behind the German lines. Numerous huts were erected on the banks of the stream, many vestiges of which still remain.

*After crossing the Rupt-de-Mad in Thiaucourt, keep along the street which continues the bridge and rises to the end of the town. On the right, towards the last of the houses, is I.C. 13, leading to* **Jaulny**.

*Recross the Rupt-de-Mad at the entrance to Jaulny. Take to the left, along*

THIAUCOURT.   BRIDGE OVER THE RUPT-DE-MAD

*the river, the road running through the village, many of whose houses were
damaged by shells.*

*On leaving Jaulny there is a large German cemetery on the right.*

*I.C. 13 runs through a pretty valley, alongside the Rupt-de-Mad and passes
near the railway station.   The old road having been destroyed by the explosion
of a German ammunition train, a new road enables the tourist, by crossing the
river, to reach the village of* **Rembercourt** *on the left bank.   The bridge was
blown up and many of the houses are in ruins.*

*I.C. 13, which continues alongside the Rupt-de-Mad as far as the Moselle,
is next reached.   This road is extremely picturesque.*

AMERICAN CEMETERY AT THIAUCOURT, AT THE SIDE OF D. 15,
HALF A MILE FROM THE VILLAGE

*Leaving Villecey-sur-Mad, slightly damaged, on the right, go towards*
**Onville**. *Three hundred yards this side of the village, to the left of I.C. 13, is*
a large German cemetery. The village, on which a few shells fell, contains a
fine church.

**Vandelainville,** *which is the continuation of Onville,* contains several
houses damaged by the bombardments.

*Passing through the villages of* **Bayonville** *and* **Arnaville,** *N. 52 bis, which*
*runs along the left bank of the Moselle, and which take to the right in the*
*direction of* **Pagny-sur-Moselle.**

This village, also called Pagny-sous-Prény, from the name of the hamlet
and château which dominates the surrounding country, was for forty-eight
years the Custom House, being the last French railway station before the
frontier.

**Prény Château,** the ruins of which are visible from here, was one of the
most famous castles of the Middle Ages. Built by the Dukes of Lorraine,

VILLAGE OF PAGNY-SUR-MOSELLE, NEAR THE CHURCH

it was dismantled by Richelieu. It formed a square flanked by high, strong
towers connected with one another by walls and subterranean passages
hollowed out of the rock. At one end there was a second building, also
surrounded by moats and flanked by towers, in one of which was the famous
"Mande-Guerre" bell. The keep with the chapel and living-rooms stood
there.

Pagny suffered severely, most of the houses being in ruins.

*The road turns to the left into the valley, then to the right beyond the church,*
*which is left on the right.*

*Just outside Pagny-sur-Moselle* the Germans built a concrete barrier across
the road to stop the tanks.

*About 500 yards from Pagny, near the bridge over Mouton stream, is a*
machine-gun blockhouse in concrete *on the right.*

*N. 52 bis next passes through* **Vandieres.** which was burnt down by the
Germans during their retreat of September 16. All the houses along the road,
especially those on the left, are in ruins.

*Five kilometres further on, after crossing the railway,* **Pont-à-Mousson** *is*
*entered by the Rue du Port, Place Colombe and Rue St. Laurent ; the latter*
*brings the tourist to the Grand Place or Place Duroc.*

PONT-À-MOUSSON.   THE BANKS OF THE MOSELLE

## PONT-À-MOUSSON

### Origin and Chief Historical Events

Pont-à-Mousson is an old town, in whose archives are found deeds dating back to 896 and 905. At that time it was called "*Villa Pontus sub castro Montionis*" (The Town of the Bridge under the Castle of Monçon).

In the 16th century there was a long controversy between the professors of the University and those of the Jesuit College as to whether the town ("Pont") or the castle ("Monçon" or "Mousson") should have precedence, *i.e.* if one should say "*Ponti Mussum*" or "*Mussi Pontum.*" The dispute was settled and the name "*Ponti Mussum*" (Pont-à-Mousson) decreed by Duke Charles III. Nevertheless, the inhabitants still insist on calling themselves "*Mussipontins.*"

Renaud I., Count of Bar, living a retired life in his château of Mousson, founded near the town in 1106 a priory dedicated to St. Michel, which he gave to the Abbey of St. Mihiel. In 1239 the "*Messins*" (inhabitants of Metz) broke down the bridge to prevent the Count of Bar communicating with his castle, but three years later they joined the Count of Bar against Duke Mathieu who, in revenge, burned down the little town of Pont.

Enfranchised in 1263 by Count Thiébaut II., Pont-à-Mousson was raised to a marquisate in 1355 by Emperor Charles IV., and, in 1356, was granted the rights and privileges of an imperial town.

Charles-the-Bold took possession of it in 1476, but it was retaken later by Duke René. However, the defection of his Swiss troops forced him to surrender it again to the Duke of Burgundy.

What made the glory and prosperity of the town was the foundation of a University in 1572. The influx of students and the renown of the professors made Pont-à-Mousson famous until 1763, when the University was transferred to Nancy.

The University encouraged the establishment of printing works, and volumes printed by Marchand and Melchior Bernard are still justly prized.

Although an open town, Pont-à-Mousson was violently bombarded by the Germans as early as August 11, 1914. After a short occupation the town was liberated by the French on September 13, 1914. The bombardments were resumed and lasted till the end of the war.

As to the part played by Pont-à-Mousson in the **Battle for the Grand Couronné Heights**, see the Michelin Guide, **"Nancy and the Grand Couronné."**

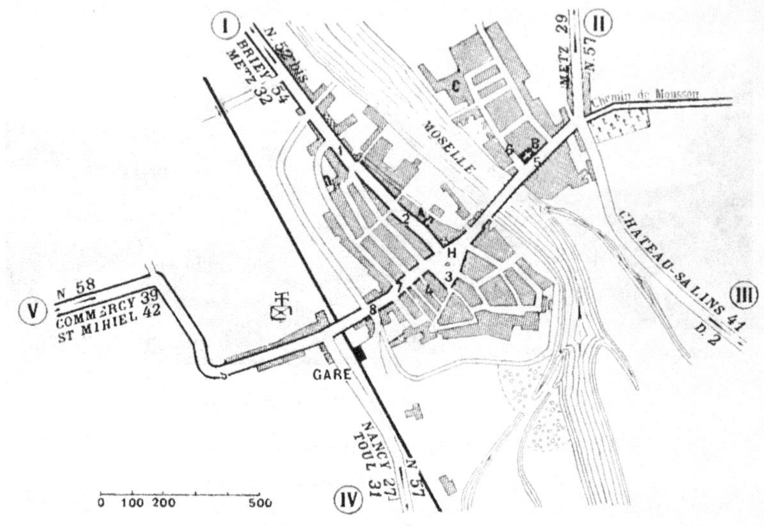



PLAN OF PONT-À-MOUSSON

*Arbitrary Signs*

A.— Church of St. Laurent.
B.— Church of St. Martin.
C.— Lesser Seminary.
H. Hôtel-de-Ville.

1.—Rue du Port.
2.—Rue St. Laurent.

3. Place Duroc.
4.—Rue de l'Union.
5.—Rue Gambetta.
6.—Rue St. Martin.
7.—Rue Victor Hugo.
8. Place Thiers.
9.—Avenue Carnot.

# A VISIT TO PONT-À-MOUSSON.

**Starting-point :** *Place Duroc (or Grand Place), in which stands the Hôtel-de-Ville.*

Place Duroc, with its irregular arcades and Renaissance houses, presents a very characteristic appearance.

*Visit first the* **"House of the Seven Capital Sins,"** decorated with

PLACE DUROC, WITH ITS ARCADED HOUSES Maison Leguy: *the 3 first arcades, beginning at the turret. The 4th and 5th arcades belong to the* House of the Seven Capital Sins.

caryatids. *At the bottom of a court there is* a fine bas-relief *representing The Conversion of St. Paul.*

*A little further on is the* **Maison Leguy**, recognisable by its hexagonal turret, which rests on one of the corner pillars of the arcades. Tradition attributes its construction to the Templars, and says that a subterranean passage led from it across the Moselle and up to Mousson.

*The turret is at the corner of Rue Victor-Hugo, which take to see, in the Rue de l'Union (first street on the left), two curious doors at Nos. 6 and 8.*

*Return to Place Duroc, turn to the left into Rue St. Laurent and see, in the court of No. 9, a fine gallery delicately carved in Renaissance style (slightly damaged by the bombardment); at No. 11, a handsome façade with a charming court, an old well, the railing of the old terrace, a spiral staircase and timbered ceilings; at No. 19, the façade, door and entrance.*

*Opposite No. 19 is the* **Church of St. Laurent;** it was slightly damaged, the roof being pierced by shells in several places.

This church, frequently restored, offers no particular interest.

Inside there is a **reredos** in the form of a tryptich, which came from the neighbouring Convent of the Poor Claires at Pont-à-Mousson. This work is by "*Georgin le painctre,*" and dates from the 16th century.

It represents : the *Baptism of Christ,* the *Resurrection of Lazarus, the Healing of the Blind at the Pool,* and the *Burial of Christ.*

The Chapel of Our Lady of Pity contains a celebrated "Christ carrying the Cross" by *Ligier Richier. (See note, p. 56, regarding this famous sculptor's works.)*

DESTROYED STONE BRIDGE OVER THE MOSELLE, WITH TEMPORARY WOODEN
FOOT-BRIDGE

CHURCH OF
ST. MARTIN.
WEST FRONT

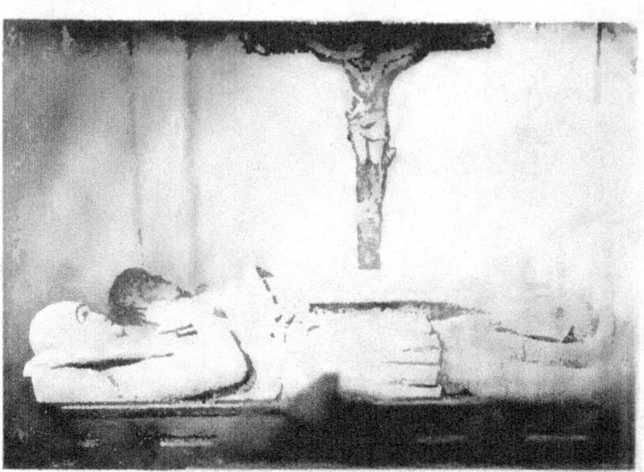



*Return to the Place Duroc, take the street leading to the bridge, which comes out opposite the "House of the Seven Capital Sins."* The fine stone bridge was partly destroyed. A temporary footway, however, makes it possible to *cross the Moselle here and reach the Rue Gambetta.*

*On the left, at the corner of Rue St. Martin, stands* the greatly damaged **Church of St. Martin.** All the stained-glass windows were destroyed. Several shells pierced the walls and roof.

The Church of St. Martin (*Hist. Mon.*) is the old church of the Antonists, and was built in 1474.

THE SEPULCHRE (16TH CENTURY) IN THE CHURCH OF ST. MARTIN

The very narrow façade is in florid pointed style. The interior of the church has undergone numerous unfortunate restorations. In the aisles are : *on the left*, the funeral statues of a 14th century knight and his wife ; *on the right*, the tomb of Esther of Apremont, with her coat-of-arms (1592), and a particularly interesting late 16th century sepulchre (*see photos. p.* 94).

A triforium runs round the nave, and the tribune is closed by a fine open-work gallery dating from the end of the 16th century. Unfortunately, the choir is disfigured by a facing of marble which conceals the frescoes that decorated the walls.

THE LIBRARY OF THE PETIT SÉMINAIRE

SPIRAL STAIRCASE
OF THE PETIT
SÉMINAIRE

*Beside the church, in the Rue St. Martin, is the* **Petit Séminaire,** housed in the sumptuous Abbey of the Premonstrants, dating from the early part of the 18th century.

It was very seriously damaged by the bombardments. The chapel and

REFECTORY
OF THE
PETIT
SÉMINAIRE

NARROW RISING ROAD TO MOUSSON, WITH SHELTER IN
THE FOREGROUND

its façade, the parlour, in very outlined rock-work style, splendid staircases, large cloisters and, above all, the famous wood-carvings in the library, were especially noteworthy.

## Mousson

*To reach Mousson, proceed to the end of the Rue Gambetta, in the opposite direction to the Moselle.*

*Leave the car at the entrance to the Avenue de Metz (on the left) and walk*

CEMETERY OF PONT-À-MOUSSON
*At the side of the above road (continuation of Rue Gambetta).*

RUINS OF TOMBS IN THE CEMETERY AT PONT-À-MOUSSON

*up the hollow road (opposite the Rue Gambetta), which skirts the cemetery. (Time required : half an hour.)*

Along this road artillery batteries were posted, the emplacements of which may still be seen.

*Take a glance at* the cemetery, where a number of graves have been destroyed.

The village of Mousson is at the top of a hill where there have been successively : a Roman camp, an Austrasian fortress and, in the 10th century, the château of the Countess Sophie de Bar, reduced to ruins by *Richelieu.*

The fortifications consist of a first-line covering the village, and a second surrounding the château. The houses thus form a semicircle between the two ramparts.

The village was greatly damaged during the war. Most of the houses are in ruins. Some of them had tricusped windows and curious 15th and 16th century doors.

*Skirt the* ancient Chapel of the Templars *to reach the* terrace of the old château.

All that remains of the château is the central chapel (11th-12th century)

A CORNER OF MOUSSON VILLAGE
*In the background :* JOAN-OF-ARC TOWER AND STATUE

PONT-À-MOUSSON AND PRÊTRE WOOD, SEEN FROM MOUSSON

(*Hist. Mon.*), which was unhappily enlarged about 1895, and to which a battle-mented tower surmounted by a gilt statue of Joan-of-Arc was added.

The chapel (*Hist. Mon.*), with a semicircular vaulted roof on curious pillars, contains fine baptismal fonts (1085) decorated with sculptures.

These fonts, resembling the curb-stone of a well, are decorated with bas-reliefs representing: *John the Baptist preaching repentance to publicans and soldiers who came to him in the wilderness ; John baptizing two naked Jews immersed in a cistern; John baptizing Jesus Christ, plunged up to the waist in the waters of Jordan.*

On the terrace are numerous trenches, in addition to shafts dug by the engineers to reach the underground passages which communicate with concrete shelters. One of these shelters may still be seen along the southern ramparts of the old château. All these military works are very interesting to visit.

There is a **splendid panorama** from this terrace : *on one side (photo, pp. 100 and 101)* the town of Pont-à-Mousson and the valley of the Moselle with, behind Pont-à-Mousson, Puvenelle Forest and Prêtre Wood ; *on the other side,* the valley of the Seille, with Metz Cathedral in the distance. *To the south-east is seen* the Grand Couronné.

Mousson was a first-rate observation-post for the French gunners, which explains the fortifications that were erected there during the war.

PANORAMIC VIEW OF PONT-À-MOUSSON AND T

RUINS OF THE OLD FORTIFIED CASTLE OF MOUSSON
*On the right :* JOAN-OF-ARC TOWER AND STATUE

...LEY OF THE MOSELLE, SEEN FROM MOUSSON

MOUSSON CEMETERY
*In the background :* WALLS OF THE OLD FORTIFIED CASTLE

# A VISIT TO PRÊTRE WOOD

**A.** *From* **Pont-à-Mousson** *to the* **Croix des Carmes,** *via* **Montauville,** *returning to* **Pont-à-Mousson**

### The Fighting in Prêtre Wood

Prêtre Wood dominates all the southern part of the Plain of Woëvre (altitude : 1,200 feet).

From October, 1914, to May, 1915, it was the scene of a continual struggle, at the end of which the wood remained in the hands of the French.

It was in September, 1914, that the Germans installed themselves in Prêtre Wood, which they at once fortified with barbed wire, chevaux-de-frise, etc.

On September 30, 1914, the French obtained a footing in the south-western edges of the forest. A month later (October 29) they captured a German post in the south-eastern salient. Their efforts were next concentrated on Père Hilarion Ravine, which they gradually occupied after many fights in the rain and snow of November and December.

Their troops advanced by short rushes as far as the principal line, which had to be taken by a direct attack. First, artillery was brought up by night to prepare the attack. Sappers, by long and patient sapping, blew up the minor defences and penetrated the blockhouses. The adversaries were at times less than a hundred yards apart.

PRÊTRE WOOD. SHELTERS IN CARRIÈRES RAVINE

From January, 1915, the French operations were directed against the western portion, towards Quart-en-Réserve and Croix des Carmes Hill. Four lines of trenches bristling with machine-guns and defences held up the attack. The ground had to be taken bit by bit, and often a counter-attack would win back in the evening the gains of several days' hard fighting. The first line was carried on January 17, and the second on February 16. At this point aerial torpedoes and hand grenades caused progress to slow down. The third line was captured on March 30. Attacks and counter-attacks followed. Fighting with hand grenades took place in the communicating trenches, behind barrages, and the artillery on both sides covered this narrow strip of ground with projectiles, breaking down the parapets and destroying the communicating trenches. The Germans, who lost heavily, brought up endless reinforcements

CEMETERY IN PRÊTRE WOOD

—in all about sixteen battalions—thus showing the importance which they attached to this position.

The final attack was launched on May 12. The French carried the blockhouses and the northern slopes beyond the crest, but the enemy still clung to the eastern and western slopes. However, the wood was won, and the splendid observation-post which the hill afforded was thenceforth in the hands of the French.

In the little cemetery on the hillside hundreds of heroes sleep their last sleep.

The slopes near the road throughout this district are one vast cemetery, while the wood proper hides beneath its soil hundreds of dead entombed by the explosion of mines or the falling-in of trenches.

This wood of tragic memories was called by the Germans "*The Wood of Death*," or "*The Widows' Wood*."

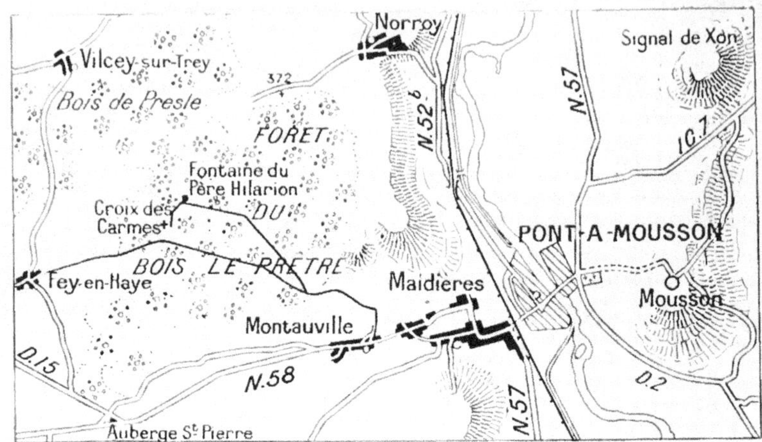


*Leave* **Pont-à-Mousson** *by Avenue Carnot, cross the railway (l.c.), leaving N.57 on the left (which follows the railway towards Nancy) and continue along N.58 to* **Montauville,** *2 km. from Pont-à-Mousson.*

This village did not suffer much. *On entering, there are several large concrete machine-gun blockhouses on the right.*

The nearest dressing station was at Montauville, in the cellar of a ruined house. First aid was given in the trenches or in the little hut near the big oak tree. From Montauville the wounded were taken in motors to Pont-à-Mousson. There was a constant procession of ambulances, stretcher-bearers and hospital attendants on the road.

*Beyond a knoll opposite the church of Montauville, take on the right a downhill road which turns sharply and leads to the village cemetery. Here the road forks. Take the road on the left. which first dips and then a little further on*

**ENTRANCE TO PRÊTRE WOOD**

*Motors stop at the fork. The road to the left leads to Fey-en-Haye,
Tourists should take the one to the right leading to the Père
Hilarion Fountain and the Croix des Carmes.*

FRENCH AND AMERICAN GRAVES ON THE ROAD TO
THE PÈRE HILARION FOUNTAIN

*rises in the direction of* **Prêtre Wood.** This road is in bad condition, but in
dry weather motors can go as far as the entrance to the wood.

*About 800 yards from the cemetery the road branches, that on the left going
to* Fey-en-Haye. *Take the road on the right, which soon leads to* **Prêtre Wood.**

*On the right, at the roadside, about 1.200 yards from the fork, are two
graves :* one of an American, the other of a French soldier.

*Three hundred yards further on, in a ravine to the right of the road, are the*
**fountain** *and* **house** *of* **Father Hilarion.** All around are numerous trenches,
shelters and military works of all kinds.

Père Hilarion fountain remained for some time between the opposing lines.
Germans and French alike came there every day to draw water, and by a tacit
understanding each side came at the definite hour. During this respite no
shot was fired from the trenches.

*Follow the road (leaving on the right a steep uphill road, 150 yards from
the fountain) which for 800 yards, down a gentle slope, crosses the* part of the

THE PÈRE HILARION FOUNTAIN AND HOUSE

THE HOUSE OF FATHER HILARION IN PRÊTRE WOOD

wood called the **"Mouchoir"** and **"Croix des Carmes"** Sector. This formed the first Franco-German lines.

The sight is a moving one : destroyed trees, the ground torn up by shells, trenches fallen in and battered shelters.

*When the road reaches the crest, look back.* The sight is more tragic still. In the distance is seen Mousson Crest, which stands out above the trees of Prêtre Wood and, *to the right, on a small hill, 100 yards from the road,* the site of the famous **"Croix des Carmes."**

GERMAN TRENCHES IN "MOUCHOIR" SECTOR, PRÊTRE
WOOD, NEAR PÈRE HILARION FOUNTAIN

PRÊTRE WOOD, THE "PELLEMENT" TRENCHES IN
"MOUCHOIR" SECTOR

When the position was taken by the French, sappers of the Engineers Corps piously removed the cross from its place and carried it to the cemetery in the valley where the heroes of these battles lie buried. There they erected it, and surrounded it with some of the barbed wire from the late German trenches.

*Return to Montauville, then to Pont-à-Mousson by the same road.*

PRÊTRE WOOD, "MOUCHOIR" SECTOR
*Gen. Le Bocq in a Trench, twenty yards from the enemy lines.*

**PRÊTRE WOOD. CROIX-DES-CARMES SECTOR**

**PRÊTRE WOOD. BETWEEN THE FRENCH AND GERMAN LINES**
*Seen from near the Croix-des-Carmes, on the road to Montauville (in the foreground).*

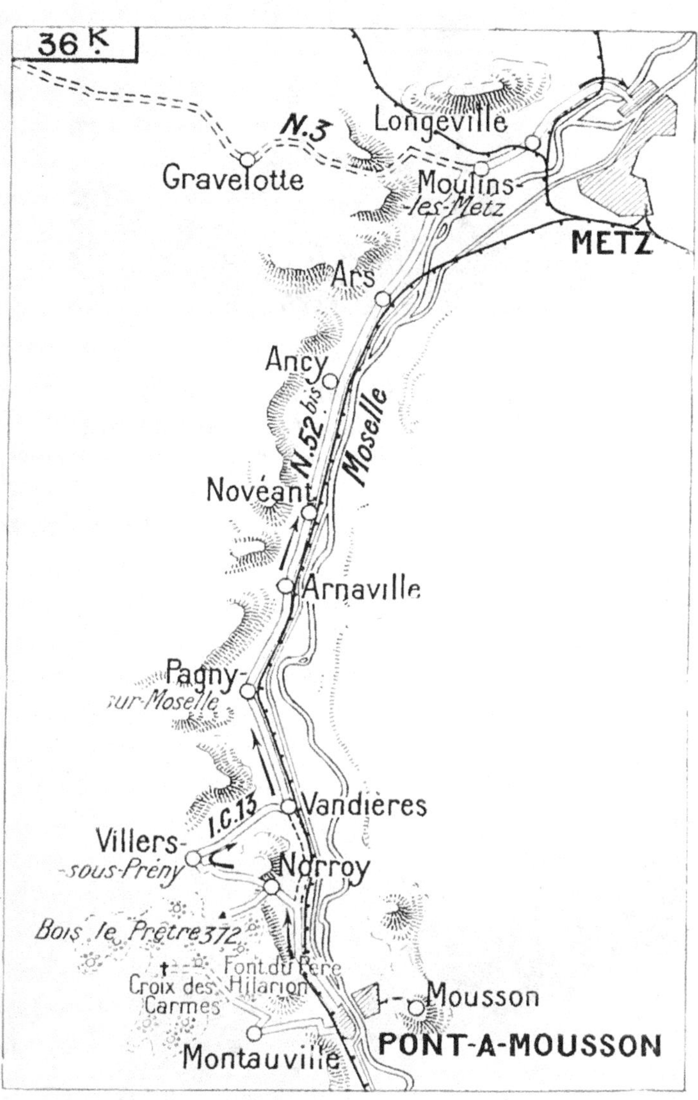


**SECOND DAY** *(continued)*

**B.** *From* **Pont-à-Mousson** *to* **Metz**

## SECOND DAY *(continued)*

### B.—PONT-A-MOUSSON TO METZ

#### From Pont-à-Mousson to Norroy and Hill 372

*At Pont-à-Mousson, on returning from Montauville, cross the railway (l.c.), take Avenue Carnot, then Rue Victor Hugo to Place Duroc. Turn to the left into Rue St. Laurent, which leads to N. 52 bis.*

*Follow the latter 3 km. to a narrow road on the left leading to* **Norroy.** *Cross*

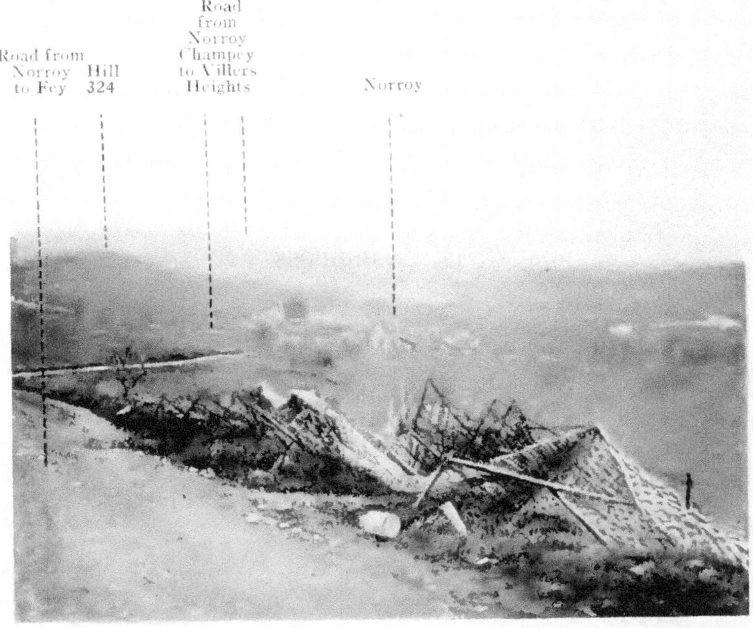


NORROY AND THE MOSELLE VALLEY, SEEN FROM HILL 372 TO THE S. W. OF NORROY, ON THE NORROY-FEY ROAD

*the village to the Place de l'Eglise. Leave the church on the right and keep along the road which rises sharply towards the crest of* **Hill 372.** *The entrance to Prêtre Wood, on the German side, is here.*

On this crest several fortified quarries served as shelters for the guns. In the wood are a number of concrete shelters, trenches and observation-posts, one of which, cupola-shaped, is well worth a visit.

*Return to Norroy, then, in front of the church, take on the left the road towards* **Villers-sous-Prény,** *which winds round Hill 372.*

GERMAN OBSERVATION-POST ON THE TOP OF HILL 372. ENTRANCE TO PRÊTRE WOOD (*coming from Norroy*)

*On leaving Norroy, the road rises sharply, then zigzags down the side of Hill 372. One kilometre from Norroy, and 100 yards to the left of the road, is a veritable village of concrete and stone, built in a quarry by the Germans, with shelters in the rock more than thirty feet deep. It served as a Post of Commandment, and was fitted with a telephone exchange which directed the artillery-fire in the Prêtre Wood sector.*

*The road continues to descend to Villers-sous-Prény (2 km.). There is a German cemetery on the left, before entering the village, many houses of which are in ruins.*

*At* **Villers** *take I. C. 13 on the right to* **Vandières** *(3 km.), where N. 52 bis is joined. Take same on the left to* **Metz.**

This road, which runs alongside the Moselle, is picturesque, but in bad condition, especially between Arnaville and Metz. (*N. 57 from Pont-à-Mousson to Metz, on the right bank of the Moselle, is in much better condition, but less picturesque.*)

*The road passes through* **Pagny-sur-Moselle** (*see* p. 109).

The valley of the Moselle becomes prettier and prettier ; varied scenery, picturesque landscapes and villages nestling in the sides of the hills. *The road turns to the left and crosses the Rupt-de-Mad stream at* **Arnaville**—the last village on the frontier since 1870, and the boundary of the old "*département*" of the Meuse.

It next passes through Novéant, where for a long time the German custom-house was installed. The village contains a château ; in the church there is a carved ivory figure of Christ.

GERMAN POST OF COMMANDMENT BELOW HILL 372
(*1 km. from Norroy on the left of Norroy-Villers road.*)

*After passing through* **Dornot** *and* **Ancy**, *the tourist soon reaches* **Ars-sur-Moselle.**

The name "*Ars*" (*Arches*) is derived from the arcades of the Roman aqueduct, the imposing remains of which are still to be seen. Known locally as the "*Devil's Bridge*," it extended as far as the village of Jouy on the right bank of the Moselle, and served to bring water to the baths and swimming-pool of the amphitheatre of the ancient *Divodurum* (Metz). It was 3,240 feet long, and 50 feet high. The church, burned down in 1807, was rebuilt in 1816 on the site of an ancient Roman fortress. *Ars* contains important ironworks and a paper factory.

**Moulins** *and* **Longville** *are next passed, after which* **Metz** *is entered by France Gate. Take the Rue de Paris, Ponts des Morts, Rue du Pont des Morts, Pont Moyen, Rue St. Marie, Rue du Faisan, Place de Chambre, then Rue d'Estrées on the right, to Place d'Armes, in which stands the* **Cathedral.**

METZ SEEN FROM THE FORT OF ST. QUENTIN

# METZ

## ORIGIN AND CHIEF HISTORICAL FACTS

The origin of Metz dates back to the Celtic epoch, when it was the capital of the *Mediomatrici*. The Romans fortified it, to defend the frontiers of the empire against the barbarians. Metz then became the centre of six great Roman roads leading to distant provinces : two from Metz to Rheims, two from Metz to Trèves (one on the right, the other on the left bank of the Moselle), one from Metz to Strasburg, and one from Metz to Mainz.

A very rich and populous town, it was embellished by numerous Roman buildings, of which excavations have laid bare important remains : an **amphitheatre,** near Porte Mazelle, and above all **Groze Aqueduct** (4th century), more than thirteen miles in length, which brought water from Gorze to Metz. Some fine remains of the aqueduct may still be seen at Jouy-aux-Arches.

The Roman Emperors who visited Metz stayed at the Governors' Palace, which stood in Place St. Croix.

Metz was taken and laid waste by the Huns in 451.

Half a century later it was rebuilt and, on the death of Clovis (511), became the capital of Austrasia and the cradle of the Carolingian dynasty. Louis-le-Débonnaire was buried in the Abbey of St. Arnoul. The Treaty of Verdun (843) gave it to Lothaire, who made it the capital of his kingdom Lotharingia (afterwards Lorraine). Thirty years later the Treaty of Mersen (870) handed it over to Louis the Germanic.

It was governed, in the name of the emperor, first by the counts and later by the bishops. In 1220, on the death of Count Thiébault, the town became a sort of republic under the title of " Free Imperial Town," and was governed by the sheriffs until 1552.

Under Henri II, the French, led by Montmorency, occupied the town, after a treaty concluded with Maurice of Saxony. The Duke of Guise, appointed Governor, energetically defended Metz, besieged by Emperor Charles-Quint (October 19, 1552). On January 1, 1553, Charles-Quint raised

111

the siege, after having lost 30,000 men. For a long time the kings of France bore the title of "Protector." Henri III. was the first to call himself "Sovereign Ruler." The Parliament of Metz, created in 1633, completed the ruin of its municipal independence, and the Treaty of Westphalia (1648) definitely incorporated it with France. It was the capital of the "Three Bishoprics" formed by the union of Metz, Toul and Verdun.

Until the Revolution (1789) Metz, while escaping the horrors of war, constantly felt its effects. Troops were continually passing through it, and its barracks became a mustering-ground. Turenne, Villars, the Marquis de Créquy, and Marshal de Villeroy camped within its walls, and it was at Metz that in August, 1744, Louis XV. was taken seriously ill, on which occasion the whole of France prayed and fasted for their "well-beloved" King.

In 1790, Metz became the chief town of the new "Département" of Moselle. Two sieges, in 1814 and 1815, were victoriously resisted.

1870 was a black year in the annals of the town—till then known as "*Virgin Metz*."* The battles of Borny (August 14), Rézonville (August 15), St. Privat (August 18), forced *Marshal Bazaine* to retire under the walls of the town. He resisted feebly, contenting himself with awaiting events, and did not even attempt to cut his way through, which would have saved the honour of the armies under his command. On October 28 he signed the capitulation, and on the following day surrendered with 173,000 men, 60 generals, 6,000 officers, 58 standards, 622 field-guns, 876 siege-guns, 72 machine-guns, 260,000 rifles and huge quantities of stores and munitions. Six months later (May 10, 1871), by the Treaty of Frankfort, Metz and part of the "*département*" of Moselle were ceded to Germany. Metz thus became the capital of German Lorraine.

It was from Metz that **La Fayette** set out in 1775 on his immortal expedition to help America win her freedom and independence. In grateful remembrance of that glorious event the **" Knights of Columbus "** recently decided to erect a statue of La Fayette in Metz (*to be inaugurated in 1920*).

GENERAL POST-OFFICE AND RAILWAY STATION

* Its coat-of-arms consists of an escutcheon argent and sable surmounted by a maiden crowned by towers and holding a palm in her left hand. It was, in fact, the proudest claim of Metz, until 1870, that it had never been taken since it had become a fortified city. In 1815 the armies of the "Holy Alliance" were refused permission to march through, when they evacuated French territory, and were obliged to cross the Moselle over a bridge which the people of Metz erected at the very foot of the ramparts, just outside the town.

## The Fortifications

From its position Metz was destined to become a stronghold of the first importance. The Romans fortified the town built by the Gauls, and erected the first citadel. The walls were preserved for a long time, and Bishop Robert, in the 10th century, utilised their remains. It was only in the 12th century that the new ramparts included the island formed by the two arms of the Moselle. They consisted of a high wall protected by sixty-eight towers. In 1552 the Duke of Guise commissioned an engineer, *Pierre Strozzi*, to restore these fortifications, which had withstood two sieges (1444 and 1552), and were in a dilapidated condition. Four years later (1556) Marshal de Vieilleville erected a citadel flanked by four bastions, on the site of the old convents. This citadel (which remained standing until 1802) stood on the site of the present Esplanade.

About a century later *Vauban*, fully aware of the strategic value of Metz, conceived a great scheme, which was carried out in the 18th century by an engineer, *M. Cormontaigne*. Vauban, for his part, added eleven new bastions to those which already guarded the citadel, but it was *Cormontaigne* who perfected the plans for inundating the valley of the Seille by utilising the waters of Lindre Pond.

Metz became finally one of the most formidable fortresses of Europe.

Under Louis-Philippe the fortifications were entirely restored, and in 1866 preparations were made to rebuild them on a new plan, better adapted to the exigencies of modern armaments and technique. Of the four detached forts of St. Quentin, Plappeville, Queuleu and St. Julien, only the first two were completed in 1870.

The Germans considerably strengthened the fortifications by means of nineteen bastions surrounded by moats, the latter being protected by thirteen out-works. The length of the line of forts was increased to eighteen miles, and eleven new forts were added.

METZ. ST. QUENTIN FORT (*seen from the Esplanade*)

METZ. AFTER THE ARMISTICE. ENTRY OF FRENCH TROOPS
*November 19, 1918.*

## METZ AFTER THE SIGNING OF THE ARMISTICE

When the Armistice was signed on November 11, the forts of Metz were within range of the American artillery, which had already bombarded them several times, while the troops had taken up the positions from which the offensive, arranged for the 16th, was to have been launched. The terms of

THE FIRST FRENCH NEWSPAPERS TO ARRIVE
*November 19, 1918.*

METZ.  FRENCH TROOPS DEFILING BEFORE MARSHAL PÉTAIN
*November 19, 1918*

PLACE D'ARMES, NOVEMBER 19, 1918
*In the background :* Statue of Marshal Fabert.

the Armistice called for the evacuation of the invaded territory, including Alsace and Lorraine, before the 26th. It was into Metz, freed of German soldiers, that the French troops made a solemn entry on Tuesday, November 19, 1918, amid scenes of indescribable enthusiasm.

The march past took place on the Esplanade, before General Pétain, made Marshal that morning. Mounted on a white horse and wearing his large blue coat, he had taken his stand in front of the statue of Marshal Ney. He was assisted by General Fayolle, commanding a group of armies, and by Major-General Buat. General Mangin, commanding the 10th Army, had met with an accident while riding, and his place was taken by General Leconte.

METZ. GENERAL PÉTAIN MADE MARSHAL OF FRANCE
*After the ceremony :* President Poincaré embraces Prime Minister Clémenceau.

On the same day M. Mirman, who had been appointed Commissioner of the Republic, was received by General de Maud'huy, Governor of Metz. Salvos of cannon and the ringing of the famous " Mutte " bell in the Cathedral celebrated this joyful day.

On the following Sunday, November 24, the leading men of Metz elected the new Town Council, and decided to restore the names of the streets in use prior to 1870, and to name new streets after generals and prominent men who had distinguished themselves in the Great War. The list was published in a decree dated December 7.

On Sunday, December 8, President Poincaré, accompanied by the French Prime Minister, M. Clémenceau, the Presidents of the Chambers, Ministers,

Marshals, and French and Allied Generals, proclaimed the definite return to France of the lost provinces. It was a day never to be forgotten by those who witnessed it. Young girls in the national costume of Lorraine—birthplace of the French President—marched through the streets, and flowers were showered from the windows on the procession.

In the morning there was a review on the Esplanade, and a Field-Marshal's *bâton* was presented to General Pétain. The President of the Republic opened the proceedings with an address, after which an unforeseen and touching incident occurred ; overcome with emotion, M. Poincaré and M. Clémenceau embraced each other.

METZ. LORRAINE GIRLS GROUPED AROUND THE FRENCH FLAG
*November 19, 1918.*

In the afternoon there was a reception in the Hôtel-de-Ville, at which President Poincaré summed up in a stirring speech the whole history of Metz, and concluded with the following words :—

"Years have gone by, but Metz has not changed. The protests formerly made to the 'Reichstag' in the name of the people of Metz, in the name of all the people of Lorraine, by that great Bishop, Mgr. Dupont des Loges, continued calmly and firmly after his death. Citizens of Metz, you renewed them, year after year, by pilgrimages to Mars-la-Tour, by visits to the cemeteries, and by fostering French memories. . . . Beloved town of Metz, your nightmare is over  France returns and opens her arms to you! "

The procession was then received with great ceremony by Mgr. Pelt at the Cathedral, and finally went to the cemetery of Chambière, to pay homage to the dead of 1870.

# PLAN OF METZ

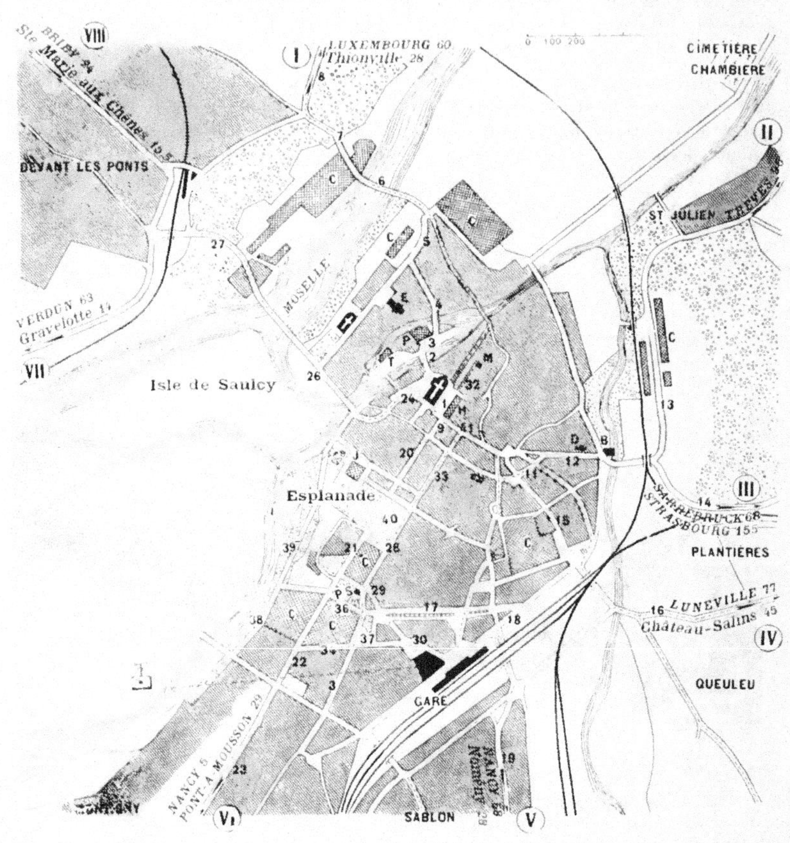


## METZ

1. Place d'Armes.
2. Prefecture Bridge.
3. Prefecture Square.
4. Rue Pont-Moreau.
5. Rue Belle-Isle.
6. Thionville Bridge.
7. Rue de l'Hôpital Militaire.
8. Route de Thionville.
9. Rue Fabert.
10. Rue de la Tête-d'Or.
11. Rue de la Grande Armée.
12. Rue des Allemands.
13. Rue de St. Julien.
14. Rue de Sarrebruck.
15. Rue Haute-Seille.
16. Rue de Strasbourg.
17. Avenue du Maréchal Foch.
18. Rue Vauban.
19. Rue de Magny.
20. Rue des Clercs.
21. Avenue de la Citadelle.
22. Rue de Nancy.
23. Rue de Pont-à-Mousson.
24. Chambre Square.
25. Rue de la Paix.
26. Rue Pont-des-Morts.
27. Rue de Paris.
28. Avenue Serpenoise.
29. Rue Harelle.
30. Rue de la Gare.
31. Rue de Thionville.
32. Rue du Haut-Poirier.
33. Rue Serpenoise.
34. Rue de Président Wilson.
35. Rue de Verdun.
36. Avenue de Maréchal Joffre.
37. King George Square.
38. Bd. Georges Clemenceau.
39. Bd. Président Poincaré.
40. Place de la République.
41. Rue Fournirue.
A. Cathedral.
B. German Gate.
C. Barracks.
D. St. Eucairés Church.
E. St. Vincent's Church.
H. Hôtel-de-Ville.
J. Palais-de-Justice.
M. Museum.
P. Prefecture.
P.S. Serpenoise Gate.
T. Theatre.

# A VISIT TO METZ

**Starting-point** : the Place d'Armes.

The Place d'Armes, in which the Cathedral and the Hôtel-de-Ville stand, is a handsome square embellished with noble buildings. On its site formerly stood the Cathedral cloister, the musicians' quarters, several chapels and private houses.

In 1753 the Governor, *Marshal de Belle-Isle,* decided that a square should be laid out there and a portal erected giving access to the Cathedral.

The plans of the architect *(Blondel)* for the portal made it necessary to lower the level of the ground. For months and years, canons and sheriffs alike stopped or impeded the work. During the night of August 9, 1755,

STATUE OF MARSHAL FABERT,
PLACE D'ARMES, METZ

M. de Belle-Isle called out the garrison, and had the work finished by torch-light. By morning the excavation was complete.

In the Place d'Armes stands a *statue of Marshal Fabert (by Etex, 1840).* The only inscription on the statue of the great Metz general (1599-1662), who was governor of Sedan, is one of his own sayings : "*If, to prevent the enemy taking a place entrusted to my care by the king, it were necessary, I should not hesitate for an instant to sacrifice myself, my family, and all my belongings.*"

METZ CATHEDRAL

## The Cathedral

The whole of one side of the Place d'Armes is occupied by the Cathedral of St. Etienne, a masterpiece of Gothic architecture. The body of the church reminds one of Amiens and Beauvais. If, on the outside, it appears somewhat narrow, the interior (393 feet long, 71 feet wide, 139 feet high), with its magnificent stained-glass, is imposing and of exceeding beauty.

The oldest portions of the Cathedral date from the 13th century.

The nave, completed in the 14th century, has eight bays. At the fourth bay it is flanked by two square towers.

The northern tower, called the "Mutte" Tower, contains the town-bell. It is surmounted by a fine spire, from which there is an extensive view of the surrounding country. It was there that the city watchman was installed, whose duty it was to give the alarm in case of fire. On the other side of the nave stands the Chapter Tower, which was finished in 1839. There is a fine doorway at the foot of each tower.

Another, smaller polygonal tower, called the Clock Tower, is built over the southern aisle.

On each side of the choir, where it meets the arms of the transept, are the two small "Charlemagne" towers, so called in memory of those which existed in the romanesque building. They give access by spiral stairways to the outside terraces over the Cathedral.

While the nave is 13th century the transept dates from the 15th, and the choir, built over a great sepulchral crypt, is contemporary with the last Gothic period.

Although completed in 1546 the Cathedral later underwent many alterations. Fires necessitated repairs, and in 1753, by order of the Governor (Marshal de Belle-Isle), the laying out of a square in front of the Cathedral necessitated the demolition of the outbuildings of the bishop's house and the erection of a portal.

INTERIOR OF METZ CATHEDRAL

The ground was excavated to a depth of some eight or nine feet, and the architect (*J. B. Blondel*) was instructed to prepare plans on a grand scale.

This was done between 1761 and 1764, after which the work was at once put in hand, and completed in 1771. While endeavouring to respect the old building, Blondel sought, not so much to build the portal in the style of the Cathedral, as to erect an independent portal in front of the church. Its irregular lines contrast with the general style of the Cathedral.

In 1791, the rood-loft, old altars and vaults were removed, in accordance with the plans of *Gardeur Lebrun*. The roof, destroyed by fire on the night of May 6, 1877—the day Emperor Wilhelm I. entered Metz—was replaced in 1880-1882 by a copper roof several yards higher than the original.

Lastly, the Doric projection of the main front was pulled down in 1903 to make room for a portal planned in the style of the rest of the church. Statues of the prophets were carved at the corners, one of which—that of the prophet Daniel—is a *likeness of the ex-Emperor of Germany, Wilhelm II.* The people of Metz would not have the ex-Kaiser-prophet take part in the entry of the French, and during the night bound his hands with a chain

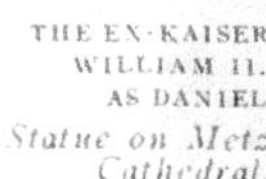

attached to which was a board bearing the inscription, "*Sic transit Gloria Mundi*" (thus passes away man's glory) (*photo above*).

The offending statue is to be replaced by a work of the Metz sculptor *Hannaux*, who designed the French monument at Noisseville.

In no other church is there so large an area of window space. It is calculated that in the transept and choir there are 4,071 square metres of glass and it is no exaggeration to say that the whole building seems to be one immense window.

METZ CATHEDRAL.  WEST FRONT

METZ CATHEDRAL.   SOUTH FRONT AND PORTAL

Among the windows are several dating from the 13th century. The large rose-window at the end of the nave, which dates from the 14th century, is the work of the master-glassworker Hermann. The windows of the north transept and the Chapel of Our Lady date from the 15th century.

Those of the south transept, Chapel of St. Nicholas, choir and apse are 16th century.

The bell called "La Mutte," which hangs in the tower of the same name, did not belong to the church, but to the town. The present bell, which is rung on all special occasions, was cast in 1505. It weighs thirteen tons and, when set in motion, causes the large and small spires to rock perceptibly. It bears the following inscription :

> "Dame Mutte suis baptisée,
> De par la Cité ci-posée,
> Pour servir à la Cité
> Aux jours de grandes solennités :
> Et aussi pour crier justice,
> Prendre ban de bonne police ;
> Les contredire quand bon me semble
> Et pour convoquer gens ensemble,"

The best view of Metz and the surrounding country is to be obtained from the top of the Cathedral tower. Here one realizes the immense importance of the forts, of which the Moselle is a kind of natural moat. On the left bank the steeply rising hills form natural defences, while the lower hills on the right bank are reinforced by the line of forts. From their gleam in the distance one gets a better idea of the number of waterways which surround and run through Metz—the River Seille, the streams of St. Pierre, Noisseville, and Châtel-St.-Germain, the River Moselle (which divides), and the canal running parallel to it. Before Metz lies the large island of St. Symphorion ; then, near the Wadrineau dyke, the smaller island of Sauley. At the foot of old Metz there is yet another arm of the Moselle, which divides, forming an island, on which stand the Prefecture and Theatre. Beyond lies the large island of Chambière, recognisable by its parade-ground and cemeteries.

METZ.   PLACE D'ARMES AND HÔTEL-DE-VILLE

### The Hôtel-de-Ville

On leaving the Cathedral the tourist should next visit the Town Hall, also in the Place d'Armes (1766-1771). The architecture is simple : façade embellished with two pediments and handsome railings. A portico leads to a fine staircase. Opposite the balustrade is a bas-relief in white marble on which are engraved the famous lines of Ausonius : "*Salve magna parens frugumque virumque Mosella . . .*" ("Hail, O Moselle ! illustrious mother of fruits and of men.")

In the interior are large reception-rooms, in which the public meetings of the Academy are held. The **Academy of Metz** was founded in 1760 by Marshal de Belle-Isle under the title of "The Royal Society of Literature, Science and Art," and endowed with the sum of sixty thousand "livres." Suppressed at the Revolution, then restored on March 14, 1819, with the motto "Useful," it obtained the title of "Royal Academy" from Charles X. on September 5, 1828. It consists of thirty-six titular members, eighteen resident members, and four honorary corresponding and associate members. The Academy largely contributed to maintain French culture in Lorraine during the German annexation.

In the grand staircase there are three windows, erected in 1852, *in the middle*, the Duke of Guise after the siege of Metz ; *on the right*, Bishop Bertram of Metz ; *on the left*, Sheriff Pierre Baudoche (1464-1489).

The flag which now flies over the building is the one which was there in

PLAN OF ESPLANADE

1870, and which was carefully preserved in the Carnavalet Museum in Paris. It was restored to the Mayor of Metz by the Vice-President of the Town Council of Paris on December 25, 1918.

*Leaving the Town Hall take Rue Fabert on the left of the Place d'Armes, then its continuation (Rue des Clercs). At the end of the latter, on the left, is the Place de la République, and on the right the Esplanade.*

The fine **Promenade de l'Esplanade** served as a parade-ground for the garrison troops, who defiled along the first row of plane-trees, past the statue of Marshal Ney (*by Pètre, 1855*). Ney, Duke of Elchingen and Prince of

GROUP OF LORRAINE GIRLS AT FOOT OF<br>MARSHAL NEY'S STATUE

STATUE OF EMPEROR WILHELM I. TAKEN DOWN BY
HIS " GRATEFUL SUBJECTS " OF METZ.   IT WAS
REPLACED BY A STATUE OF " LE POILU "

Moskowa, was born at Sarrelouis.   He is represented, rifle in hand, ready
to fire.

*Go to the end of the Esplanade, beyond the bandstand on the terrace;* magnificent view of the Hill and Fort of St. Quentin, Fort Plappeville and the Moselle.   The island of Saulcy, on which stands the powder-factory, is just opposite.

It was on this terrace that the bronze equestrian state of *Kaiser Wilhelm I.*

THE " POILU " STATUE, WHICH REPLACES
THAT OF WILHELM I.

THE " PROMENADE DE LA MOSELLE "

(1892) used to stand. According to the inscription on the pedestal the statue was erected by the Emperor's "grateful people." The conqueror was represented pointing to the Moselle and the powerful forts of Plappeville and St. Quentin which protect the town.

The "grateful people" dragged this statue off its pedestal into the mud a few days before the French entered the town, and on the night of January 6 replaced it with a statue "*To the Victorious Poilu*," *bearing the inscription* "*On les a*" (variation of the famous rallying cry "*On les aura*") as a pleasant surprise for Marshal Pétain who, next day, was to decorate sixteen regiments with the "fourragère" cord and bestow decorations on various officers and soldiers. This statue was made in seven days by the local sculptor Bouchard (*photos, p.* 128).

*In the Esplanade stands the Palais-de-Justice* (1776), *on the site of the former* Hôtel de la Haute-Pierre, the property of the Duke of Suffolk, lover of Mary Tudor, Queen of England. He had this mansion pulled down and the fine Hôtel de Suffolk built, which, for a long time, served as the Government House. Finally, in 1776, *Clairisseaux* built the present palace. The **iron railings** of the grand staircase and, *in the inner court*, **two bas-reliefs** one recalling the humanity of the Duke of Guise in succouring the soldiers of the Duke of Albe after the raising of the siege ; the other glorifying the peace concluded in 1783 between England, France, Spain, the United States of America and Holland, are especially noteworthy.

STATUE OF
KAISER
FREDERICK
CHARLES
DRAGGED
DOWN FROM
ITS
PEDESTAL
BY THE
PEOPLE OF
METZ

*Return to the Place de la République and take on the right the Avenue de la Citadelle, which separates the Esplanade from the Place de la République. Follow this avenue, which soon skirts on the left the Engineers' Barracks, and a garden.*

*Beyond the garden, turn to the left into the Avenue du Maréchal Joffre, which leads to the Place du Roi-George (in front of the old railway station). It was here that the statue of Kaiser Frederick III. was pulled down by the people. Not far from this square may be seen a round tower—a relic of the ramparts of the Middle Ages.*

*Turning his back on the old railway station, the tourist next takes Avenue Serpenoise (beside the gardens), along which run the tram lines. On the left is the Serpenoise Gate (1852).*

*Continue along the Avenue, which skirts, on the left, first the Engineers' Barracks, then Place de la République.*

*Beyond the latter the Avenue is continued by Rue Serpenoise—the busiest street in Metz—which take. Rue Ladoucette, which continues it, leads to Rue Fournirue.*

*Take the latter on the right, then Rue du Change (which continues it to the right) to* **Place St. Louis.**

SERPENOISE
GATE,
*leading to
the Place de la
République.*

PLACE ST. LOUIS AND THE ARCADES

In former times, Place St. Louis (or Place du Change) was occupied by sixty moneychangers' stalls. Several of the houses in the square have retained their battlements, pointed or semicircular arches, tricusped windows and Renaissance balconies. The name of St. Louis comes from a statue of Louis XIII., found among the ruins of the citadel and which the Curé of St. Simplice took for one of Louis IX. Mystery plays used to be acted in the square, which later was used for the execution of criminals. Finally, it became the corn market.

*At the end of the square take Rue Royale, then turn to the left into Rue Coislin, which skirts the Coislin Barracks.*

*At the end of Rue Coislin take Rue Pont-à-Seille to Place des Charrons, then, at the end of this square, Rue du Grand-Wad, to the Rempart des Allemands. Follow the latter to the left as far as the* **German Gate.**

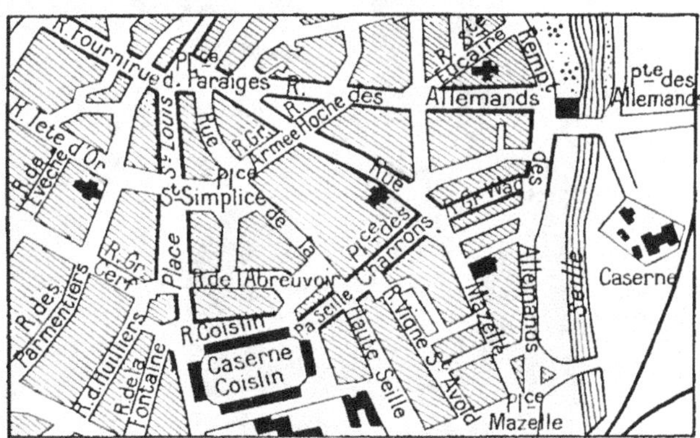

PLAN OF PLACE ST. LOUIS

PORTE DES ALLEMANDS (GERMAN GATE)
*Seen from the Quai des Allemands.*

The **German Gate,** on the banks of the Seille, is a remarkable structure.
Mention of it occurs as early as 1324. In the 15th century it was completely restored by the architect *Henri de Banceval.*

*Opposite the Gate take the Rue des Allemands ; on the right is the interesting Church of St. Rucaire. Continue to Place des Paraiges.*

THE GERMAN GATE

THE
GERMAN
GATE

*Seen from
the right
bank of
the Seille.*

THE TAN-YARDS

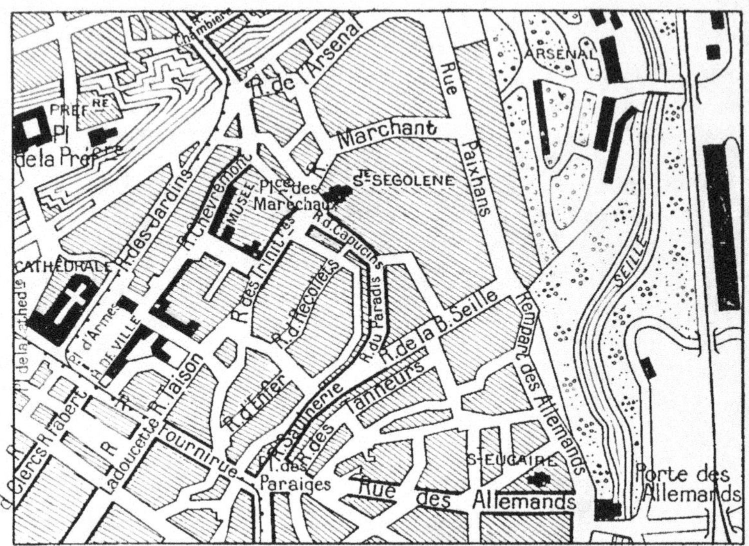


*At the end of the square, take Rue Saulnerie to the right (continued on the left by Rue du Paradis), which leads to Rue des Capucins. At the end of the latter is Place des Maréchaux, in which stands the* Church of St. Ségolène, built on the site of an oratory founded by St. Ségolène in the 8th century. The present church, built at two different periods (the choirs, nave and portal are earlier than the aisles), dates from about the 13th century. Long, narrow windows, mostly double, end in stanchions. The two side chapels contain fine stained-glass. Note the curious open-work gallery of the organ loft, and several interesting paintings.

*Turn to the left and take the Rue des Trinitaires. Skirt an old building with a square turret, beside a doorway—" Hostel St. Ligier "—then turn to the right into Rue de la Bibliothèque.*

*In this street, at the corner of Rue Chèvremont, there is* a large building (formerly the Church of the Petits-Carmes), *the work of Sébastian Leclerc, in* which are housed both the library (80,000 volumes and 1,987 manuscripts) and the Museum (local archæology, natural history, objects of art and three rooms of pictures).

*Besides the museum, take Rue Chèvremont, which runs into Rue de la Boucherie, in which turn to the left to St. Georges Bridge over the Moselle.*

*Cross this bridge, from which there is a lovely view, and take the Rue du Pont St. Georges. Rue Chambière opens at once on the right, and leads to* **Chambière Cemetery,** in which are the graves of the French soldiers who fell in the siege of 1870.

*The road passes between the large slaughter house and cattle market, and*

MOSELLE RIVER
*Seen from St. George's Bridge.*

*huts serving as an army stores.  Cross an* old cemetery, in the middle of which are several monumental tombs.  *Skirt the Jewish cemetery and the Moselle, as far as the* **Military Cemetery** : numerous graves under the trees.  In the centre stands a pyramid thirty-seven feet high, with a great number of piled up coffins carved on the base.  Here lie the soldiers who died in the Metz hospitals of wounds received in the battles of Borny, Gravelotte, St. Privat, Servigny, Peltre and Ladonchamps — 7.203 in number.

MOSELLE RIVER AND ST. GEORGE'S BRIDGE

On the principal façade is a bas-relief in white marble representing religion, taken from a disused vault belonging to the *de Salse* family. On the other side are inscriptions. At the base of the pyramid is the inscription : "*The Women of Metz to those whom they nursed.*"

Beside the pyramid there is a monument to the memory of the fallen French officers.

For forty-eight years wreaths, tri-colour cockades and ribbons were piously placed on these graves, and on each anniversary day the women of Metz covered them with flowers.

*Take Rue du Pont St. Georges to Rue St. Vincent (on the left), which follow, then turn to the right into Rue des Bénédictins.*

*Apply at No. 7 to visit the* **Church of St. Clément.**

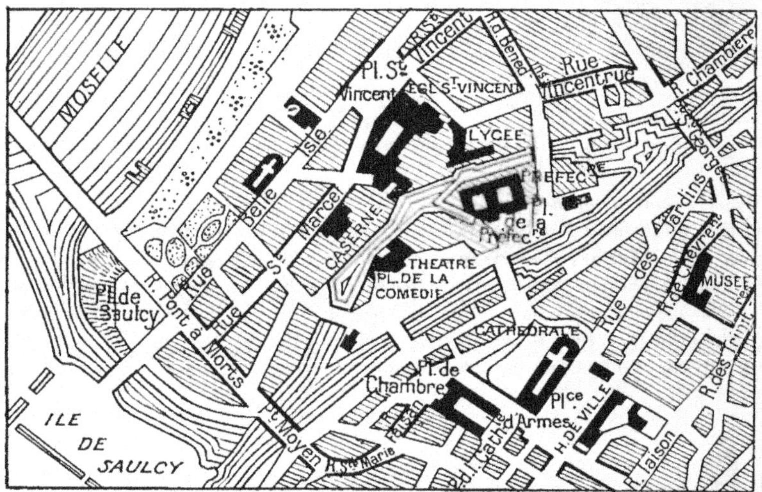


Founded in 1668, the choir, nave and aisles were begun in 1680 by *Spinga*, an Italian. The portal was damaged during the Revolution. To-day the church forms part of the college founded by the Jesuits. A fine cloister with a well should be visited.

*Return to Rue des Bénédictins and follow it as far as Rue St. Vincent (on the left), which leads to the square of the same name, where stands the curious* **Church of St. Vincent,** *founded in 1248.*

Partially destroyed by fire in 1711, by an apostate monk, it was used as a stable during the Revolution, and then as a hospital in 1814. Once more a church, a portal in composite style was added. The graceful nave on twelve shafted pillars, the symmetrical choir and the fine chapels in pointed style are well worth seeing.

*Continue along Rue St. Vincent, on the other side of the square. Its continuation, Rue St. Marcel, leads to Rue du Pont-à-Mort, into which turn to the left.*

ST. MARCEL BRIDGE AND THE PROTESTANT CHURCH
*Seen from Moyen Bridge.*

Cross the Moselle by Moyen Bridge (lovely view). *Take Rue St. Marie, which continues the bridge,* then Rue du Faisan on the left, *leading to the* pretty little Place de Chambre. This square owes its name to the Knights of Malta, who in 1323 lived there in a castle called *Petit St. Jean.*

*From the Place de Chambre return to the Cathedral and to the Place d'Armes by the narrow Rue d'Estrées.*

MOYEN BRIDGE

MOYEN BRIDGE AND THE CATHEDRAL
*Seen from Saulcy Island.*

## THIRD DAY

### METZ-ETAIN-VERDUN

(*See Itinerary, pp. 138-139*)

*Leave Metz (Place d'Armes) by Rue d'Estrées, cross Place de Chambre (soon reached on the left), take Rue Faisan, then Rue St. Marie, leading to Moyen Bridge over an arm of the Moselle. Cross the bridge and take Rue du*

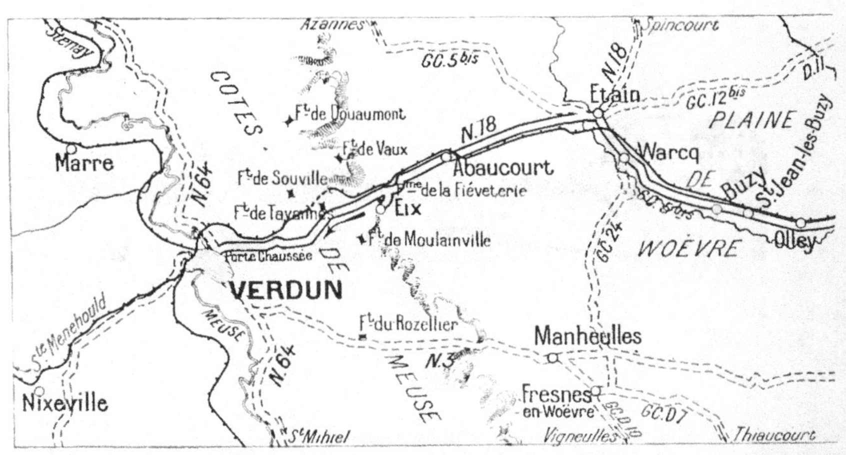

THE PREFECTURE BRIDGE

*Pont-à-Mort. Cross the ramparts, then the second arm of the Moselle. Take Rue de Paris and follow the tram lines towards Moulins ; after crossing the second belt of ramparts and the railway, the route turns to the left.*

*The road hereabouts is bordered with fine trees. After passing through* **Ban-St.-Martin** *(Infantry Barracks on the right) and* **Longeville-les-Metz,** *the tourist arrives at* **Moulins.**

*At the fork, take the Verdun road, on the right, which passes in front of the barracks.*

*Two hundred yards beyond Moulins leave on the right the Briey road, and at the milestone marked " Metz 7 km.," turn to the left into the uphill road to* **Gravelotte.**

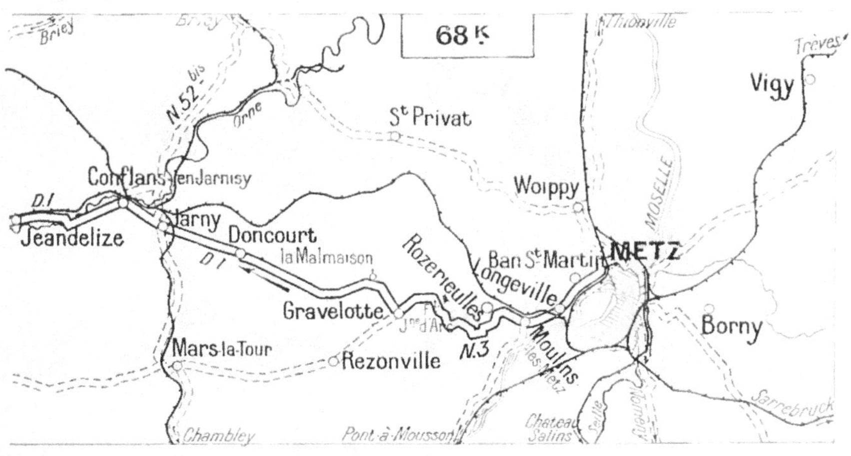

*Near milestone "Metz 9 km.," there is a fine view of Metz : in the foreground* the village of Rozérieulles is seen in the valley ; *in the background,* the Moselle valley and Metz.

*At the top of Hill 342, the road passes near* **Joan-of-Arc Fort,** formerly the German fort "*Kaiserin.*" It stands about 300 yards to the right of the road. *The latter, a little further on, turns sharply to the left near a* **monument** surrounded with trees, which was erected to the memory of the soldiers who fell in 1870. Several graves bear the inscription : "Krieger v. 18 8-1870."

*St. Hubert Farm is soon reached, then the deep* **Ravine of Mance,** *along* which the defeated Germans were forced to retreat in the course of the great battle of August 16, 1870, fought between the villages of Gravelotte (*which the tourist soon reaches*), Rezonville and Mars-la-Tour (*further west*).

In **Gravelotte** (12 *km.* 700 *from Metz) take the road to Doncourt (D.1.) on the right, in front of the Post Office. The road passes close to* **Mogador Farm,** from which Kaiser Wilhelm I. saw his troops thrown into confusion and beating a hurried retreat under the cover of night.

*After passing in front of* **Malmaison Farm,** *the old frontier is crossed.*

*Eight kilometres 900 beyond Gravelotte.* **Doncourt** *village is reached. Cross through and keep straight on along D.1. After passing by Jarny Mines, the road crosses the railway (l.c.) and enters* **Jarny** *village, 4 km. from Doncourt.* Several houses were destroyed and the church badly damaged.

*Two kilometres beyond Jarny,* **Conflans-en-Jarnizy** *is reached.* Several of the houses were destroyed. *Go through the village and at the far end take the Etain road. 5 km. from Conflans,* **Jeandelize** *is reached.* The church (*on the right*) was torn open by shells. Note the Renaissance doors of several of the houses.

*Keep straight on.* **Olley** *village (on the right) is passed, 2 km. beyond Jeandelize. There is a large German cemetery fifty yards from the road on the right.*

**St. Jean-lés-Buzy** *and* **Buzy** (*the latter 11 km. from Conflans) are passed through, after which* **Hill 198** —on which is a German stronghold with blockhouse, trenches and barbed-wire entanglements—*is reached.*

The partly-destroyed village of **Warcq** *is next passed through, after which* 2 km. further on, **Etain** *is reached.*

ETAIN. RUINED CHURCH AND HOUSES
*Seen from the bridge over the Orne, on the road to Verdun.*

# ETAIN

Etain was looted by the Swedes in 1622, during the reign of Louis XIII. Later, it was often taken and retaken by the French, Germans, Spaniards and Lorrains. Its fortifications were destroyed under Louis XIV. By the Treaty of Vienna (18th century) the town was definitely ceded to France.

In October, 1792, Kellermann's advance guards, in pursuit of the Prussians, encamped at Etain.

ETAIN. THE TOWER AND SOUTH FRONT OF THE CHURCH
*The tower was used as an observation-post by the Germans.*

ETAIN.    CHEVET AND NORTH FRONT OF CHURCH

In 1914, the town was bombarded by the Germans on August 24, from 1 p.m. to 2 a.m. the next day, and again on the 25th at 11 o'clock, with incendiary shells.

Many of the inhabitants were killed on the 24th.  On the 25th others, who had taken refuge in the cellars of the Town Hall, perished under the ruins of that building.  200 fled along the Verdun road.  A girl telephonist remained

ETAIN CHURCH.    CENTRAL NAVE SEEN FROM THE CHOIR

at her post and kept in touch with Verdun every quarter of an hour. Her last message (on the 25th) was: " A bomb has just fallen on the office."

The same day French troops routed the German XXXIIIrd D.R. in a glorious battle at Etain. Nevertheless, the enemy occupied the town, which was systematically looted. Every two days train-loads of furniture, linen, wines, food, cloth, boots, tools and raw materials were sent to Germany.

In April, 1915, French troops captured Hills 219 and 221, Hôpital Farm (formerly belonging to Order of St. Jean de Rhodes) and Haut-Bois Farm, reaching the immediate vicinity of the town, without, however, entering it.

*In the* partly destroyed town, V. 18 *is picked up again, which take to the left. The greatly damaged church (photos, pp. 141 and 142) is seen on the right.* Its belfry was torn open by the bombardments, leaving visible the interior, where the Germans had installed an observation-post.

*Viollet-le-Duc* considered the Church of Etain, with its three naves, as one of the five most remarkable churches in the Meuse province. Begun in the 13th century, it was completed in the 15th. The imposing choir, with its large, many-mullioned windows, is 15th century. In the right aisle are a remarkable holy-water basin, and a statue of Our Lady of Mercy by *Ligier Richier.* The basin is of bell-metal and, like those of Nevers and Bourges,

ETAIN CHURCH. LIGIER RICHIER'S " DESCENT FROM THE CROSS "

PANORAMIC VIEW O

FIÉVÈTERIE FARM AND VILLAGE OF EIX
*Seen from the Verdun road.*

HE MEUSE HEIGHTS

shaped like a mortar, but the epitaph round it proves its sacred origin. The *Ligier-Richier* group (1528) represents the Virgin Mary gazing on the dead body of Christ. It differs slightly from that of Clermont-en-Argonne, attributed to the same sculptor.

*Beyond Etain, N. 18 crosses the Orne stream (photo, p. 111) and the military defences of the town, of which several concrete works remain.*

*The road crosses Woëvre Plain. Shortly before reaching the level-crossing, before arriving at the railway station and village of* **Abaucourt,** the Meuse Heights can clearly be seen on the horizon, at the end of Woëvre Plain *(panorama above).*

*Go through* **Abaucourt** *(razed to the ground) to* **Fiévèterie Farm** *(in ruins), which lies at the foot of the Meuse Heights (photo. p. 111). A road starts on the left of the farm and leads to the small ruined village of* **Eix,** *which was the scene of fierce fighting throughout the war.*

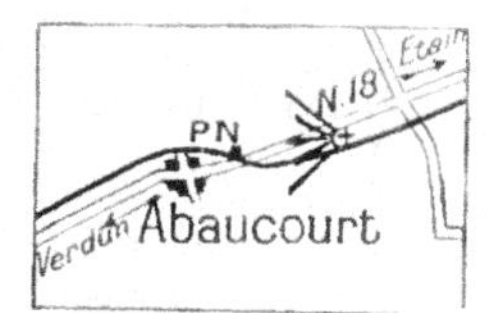



*The road up the* **Meuse Heights** *is fairly steep and passes between the* **Forts of Souville** *and* **Tavannes** *(on the right) and* **Moulainville Fort** *(on the left). It then descends in a gentle slope to Verdun, which is entered by the Faubourg Pavé and Chaussée Gate.*

# CONTENTS

MICHELIN ILLUSTRATED GUIDES
TO THE BATTLEFIELDS (1914-1918)

# THE AMERICANS
## IN THE
# GREAT WAR

### VOLUME I.
### THE SECOND BATTLE OF THE MARNE
(CHATEAU THIERRY, SOISSONS, FISMES.)

*a panoramic history and guide*

*Price* **$1**

MICHELIN & Cie., CLERMONT-FERRAND
MICHELIN TYRE Co. Ltd., 81 Fulham Road, LONDON, S. W
MICHELIN TIRE Co., MILLTOWN, N. J., U. S. A

THE AMERICANS IN THE GREAT WAR.    VOL. I.

MICHELIN ILLUSTRATED GUIDES
TO THE BATTLEFIELDS (1914-1918)

# THE AMERICANS
## IN THE
# GREAT WAR

VOLUME III.
MEUSE-ARGONNE BATTLE
(MONTFAUCON, ROMAGNE, ST MENEHOULD)

*a panoramic history and guide*

Price **$1**

MICHELIN & Cie., CLERMONT-FERRAND
MICHELIN TYRE Co. Ltd., 81 Fulham Road, LONDON, S. W.
MICHELIN TIRE Co., MILLTOWN, N. J., U. S. A.

PRINTED IN THE U. S. A
BY THE ESSEX PRESS, INC., NEWARK, N. J

# BEAUTIFUL FRANCE
### Paris and its Environs

*PARIS*—home of grandeur, elegance, and wit—plays a part in France probably unequalled in any other country, and may be considered, in many respects, as the chief city of Europe, and one of the greatest in the world. Above all, it possesses eminently national qualities which ten centuries of refinement and taste have handed down to contemporary France.

It is impossible, in a few lines, to paint the exceptional charms of Paris which the whole world admires.

Its vistas of the Champs-Elysées seen from the Tuileries and the Arc de Triomphe; of Notre-Dame and the point of the City Island seen from La Concorde Bridge; of the River Seine, the Institute, the Louvre, seen from the Pont-Neuf embankment; Notre-Dame and its quays, seen from the end of St. Louis Island; the panorama of the city seen from the top of Montmartre Hill; the Avenue du Bois de Boulogne and the Bois de Boulogne itself, etc., etc.—all are of incomparable beauty.

The city's historical monuments are of inestimable value, and the most famous art treasures are to be found in its Museums.

The surroundings of Paris join the charm of their landscapes to the world-wide fame of their parks and castles: *Versailles*, whose palace and park recall the splendour of the Louis XIV. period, and where the "Trianons" have preserved graceful traces of the Court of Marie Antoinette; *St. Germain*, with its castle and forest; *St. Cloud* and its park; *Sèvres* and its world-renowned art porcelain factory; *La Malmaison*, home of Bonaparte before he became Napoleon I.; *Rambouillet*, *Fontainebleau*, *Chartres* with its marvelous cathedral, *Maintenon*, *Dreux*, etc.—all these form a girdle round Paris such as no other metropolis in the world can boast of.

# HOTELS AND MOTOR AGENTS
### AT
# SAINT MENEHOULD, VERDUN.



Hôtel Saint Nicholas, Rue Chanzy.

Hôtel de Metz, 23, Rue Chanzy.

E. Depors, 3, Rue Chanzy.

---

## VERDUN

### Hostellerie du Coq Hardi, 8, Rue du St. Esprit
(between the Rue Mazel and the Meuse)

### Hôtel du Lion d'Or, Place Saint Paul
(Opposite the Sub-Prefecture)

---

Grand Garage Central Rochette, 22, Rue de la Rivière. Inspection pit. Petrol. Telephone No. 50.

---

The above information may no longer be exact when it meets the reader's eye. Tourists are therefore recommended to consult the Michelin Touring Office.

Before setting out on a motoring tour, whether in the British Isles or abroad, call or write to:

### THE MICHELIN TOURING OFFICE

81, Fulham Rd., London.
——S.W. 3,——

who will be pleased to furnish all desired information and a carefully worked-out itinerary of the route to be followed free of charge.

Made in the USA
Monee, IL
07 July 2026

56548190R00089